Eugen Cornelius Joseph von Lommel

The nature of Light, with a general Account of physical Optics

Eugen Cornelius Joseph von Lommel

The nature of Light, with a general Account of physical Optics

ISBN/EAN: 9783337025069

Printed in Europe, USA, Canada, Australia, Japan

Cover: Foto ©berggeist007 / pixelio.de

More available books at **www.hansebooks.com**

THE INTERNATIONAL SCIENTIFIC SERIES.

THE NATURE OF LIGHT,

WITH A GENERAL ACCOUNT OF

PHYSICAL OPTICS.

BY

Dr. EUGENE LOMMEL,

PROFESSOR OF PHYSICS IN THE UNIVERSITY OF ERLANGEN.

WITH ONE HUNDRED AND EIGHTY-EIGHT ILLUSTRATIONS AND A PLATE OF SPECTRA
IN CHROMOLITHOGRAPHY.

NEW YORK:
D. APPLETON AND COMPANY,
72 FIFTH AVENUE.
1898

PREFACE.

THE OBJECT of this little work is to give to a large circle of readers an answer, based on the present state of science, to the question, What is the Nature of Light?

In the first fourteen Chapters the laws of reflexion, refraction, dispersion, and absorption of light are demonstrated by experiment without reference to any theory of the nature of light. This comes forward prominently for the first time in the fifteenth Chapter, in discussing Fresnel's mirror experiment, and the conclusion arrived at being in favour of the undulatory theory, it is shown that this theory is not only in accordance with all the facts hitherto known, but also affords the most satisfactory explanation of the phenomena of double Refraction and polarisation, both of which receive subsequent consideration.

Mathematical reasonings are wholly omitted in the text; where these are required or appear to be desirable for the more thorough and complete knowledge of the phenomena described, they are given in the most elementary form, and are added as an appendix to the Chapters.*

Numerous wood-cuts are introduced, many of which are taken from the Atlas of Physics of Johann Müller; the majority, however, are new, as is also a chromo-lithographic plate of spectra.

I trust that this attempt to render a branch of Physics, which at first sight seems from its delicate nature to lie somewhat beyond the grasp of the general public, intelligible, will meet with a kindly reception and consideration at their hands.

ERLANGEN, *July* 1874.

---

* The theory of spherical mirrors and lenses, for example, and the elementary theory of the rainbow, are added as Appendices to the Chapters in which these subjects are discussed.

CONTENTS.

CHAPTER		PAGE
	PREFACE	v
I.	SOURCES OF LIGHT	1
II.	RECTILINEAR PROPAGATION OF LIGHT	14
III.	REFLEXION OF LIGHT	26
IV.	SPHERICAL MIRRORS	40
	APPENDIX TO THE FOURTH CHAPTER	50
V.	REFRACTION	56
	APPENDIX TO THE FIFTH CHAPTER	73
VI.	LENSES	78
	APPENDIX TO THE SIXTH CHAPTER	90
VII.	OPTICAL INSTRUMENTS	95
VIII.	DISPERSION OF COLOUR	112
	APPENDIX TO THE EIGHTH CHAPTER: THEORY OF THE RAINBOW	126
IX.	ACHROMATISM	134
	APPENDIX TO THE NINTH CHAPTER. ACHROMATIC LENSES	146

CONTENTS.

CHAPTER		PAGE
X.	SPECTRUM ANALYSIS	148
XI.	SPECTRUM ANALYSIS OF THE SUN	159
XII.	ABSORPTION	172
XIII.	FLUORESCENCE. PHOSPHORESCENCE. CHEMICAL ACTION	183
XIV.	ACTION OF HEAT	197
XV.	FRESNEL'S MIRROR EXPERIMENT: UNDULATORY MOVEMENT	207
XVI.	PRINCIPLE OF INTERFERENCE. CONSEQUENCES OF FRESNEL'S EXPERIMENT	217
XVII.	HUYGHENS' PRINCIPLE	229
XVIII.	DISPERSION OF LIGHT. ABSORPTION	242
XIX.	DIFFRACTION OF LIGHT	258
XX.	COLOURS OF THIN PLATES	273
XXI.	DOUBLE DIFFRACTION OF LIGHT	282
XXII.	POLARISATION	293
XXIII.	POLARISING APPARATUS	303
XXIV.	INTERFERENCE OWING TO DOUBLE REFRACTION	316
XXV.	CIRCULAR POLARISATION	332
	INDEX	353

LIST OF ILLUSTRATIONS.

FIG.		PAGE
1.	Bunsen's burner	3
2.	Oxygen lamp	6
3.	Drummond's light	6
4.	Magnesium light	8
5.	Electric light between carbon points	8
6.	Electric lamp	11
7.	Shadows	15
8–9	Shadow nucleus, and penumbra	16
10.	Projection of an image through a small aperture	20
11.	Diminution of illumination in the ratio of the square of the distance	22
12.	Bunsen's photometer	23
13.	Rumford's photometer	24
14.	Reflexion of light	26
15.	Model to demonstrate the law of reflexion of light	28
16.	Production of image-point in a plane mirror	29
17.	Production of the image in a mirror	30
18.	Mirror-image in a transparent plate of glass	31
19.	Heliostat	32
20.	Reusch's heliostat	33
21.	Principle of reflecting goniometer	34
22.	Angular mirror	36
23.	Principle of the mirror sextant	37
24.	Mirror of reflecting sextant	38
25.	Concave mirror	40
26.	Focus	41
27.	Conjugate foci	42
28.	Conjugate points	44
29.	Conjugate points on a secondary axis	45
30.	Real image	46
31.	Mode of production of real images	46

LIST OF ILLUSTRATIONS.

FIG.		PAGE
32.	Mode of formation of virtual image	48
33.	Virtual principal focus of a convex mirror	49
34.	Production of a virtual image behind a convex mirror	50
35.	Mode of expressing the size of any angle	50
36.	Determination of the position of the principal focus	51
37.	Determination of the position of conjugate points	52
38.	Construction showing the formation of the image	54
39.	Refractor	56
40.	Angles of incidence and of refraction	57
41.	Apparatus for demonstrating the law of refraction	58
42.	Law of refraction	60
43.	Total reflexion	61
44.	Totally reflecting prism	64
45.	Apparent position of a point situated beneath the surface of the water	65
46.	Appearance presented by a rod dipped in water	65
47.	Refraction through a transparent plate with parallel surfaces	65
48.	Refraction through two parallel plates	67
49.	Refraction through a piece of glass the surfaces of which are not parallel	68
50.	A prism	68
51.	Deflection through a prism	69
52.	Smallest deflection through a prism	70
53.	Hollow prism	71
54.	Construction of the refracted ray	73
55.	Refraction through two parallel plates	74
56.	Passage of a ray of light through a prism	76
57.	Convex lenses	78
58.	Concave lenses	78
59.	Axis and centres of curvatures	79
60.	Focal point	80
61.	Conjugate foci	81
62.	Conjugate foci	82
63.	Virtual image	82
64.	Production of a real image	84
65.	Real image seen through a convex lens	86
66.	Virtual image with a convex lens	87
67.	Virtual focus of a concave lens	88
68.	Action of a concave lens on convergent and divergent rays	89
69.	Virtual image formed by a concave lens	89
70.	Determination of the focal distance	90
71.	Determination of conjugate points	93
72.	Duboscq's lamp	95

LIST OF ILLUSTRATIONS. xi

FIG.		PAGE
73.	Magic lantern	97
74.	Solar microscope	99
75.	Camera obscura	101
76.	Action of the microscope	102
77.	Microscope	103
78.	Mode of showing the image of a microscope as an object	104
79.	Action of the astronomical telescope	105
80.	Astronomical telescope	106
81.	Instrument for measuring the prismatic deflection	106
82.	Terrestrial telescope	107
83.	Construction of Galileo's telescope	108
84.	Galileo's telescope	108
85.	Action of Newton's reflecting telescope	109
86.	Action of the reflecting telescope with front opening	110
87.	Gregory's reflecting telescope	110
88.	Action of Gregory's reflector	111
89.	Vaporisation of metal in the arc of the electric flame	112
90.	Different deflection of different coloured rays of light	115
91.	Undecomposability of the colours of the spectrum	118
92.	Impure spectrum obtained by the use of a circular opening	119
93.	Combination of the colours of a spectrum to form white light	119
94.	Complementary colours	120
95.	Combination of two homogeneous colours	121
96.	Refraction and internal reflexion in a rain-drop	123
97.	Refraction and double internal reflexion in a rain-drop	124
98.	Mode of formation of the rainbow	125
99.	Refraction and internal reflexion in a drop of water	126
100.	Theory of the rainbow	129
101.	Combination of two similar prisms without deflection and without dispersion.	136
102.	Combination of a crown and flint-glass prism causing dispersion but no deflection	137
103-4.	Combinations of prisms which cause no deflection	138
105.	Combination of a crown and flint-glass prism, with deflection, but without dispersion (an achromatic prism)	138
106.	Spectrum thrown by crown glass and by flint glass	139
107.	Dispersion of colour of a lens	140
108.	Achromatic lens	141
109.	Measurement of refraction as practised by Fraunhofer	142
110.	Spectrometer	144
111.	Bunsen's spectroscope	148
112.	Induction apparatus	154
113.	Geissler's spectrum tube	155

LIST OF ILLUSTRATIONS.

FIG.		PAGE
114.	Action of the comparison prism	159
115.	Comparing prism at the slit of the spectroscope	160
116.	Bunsen's apparatus for the absorption of Sodium light	162
117.	Absorption of the Sodium flame	163
118.	Telescope with four prisms	166
119.	Absorption spectra of nitrous oxide and of the vapour of iodine	173
120.	Absorption spectra	175
121.	Absorption of the colouring matter of litmus with different thicknesses of the layer	178
122.	Fluorescence	183
123.	Solar spectrum with the ultra-violet portion	185
124.	Geissler's fluorescence tube	188
125.	Geissler's tube with Uranium glass spheres	188
126.	Absorption and fluorescing spectrum of Naphthalin-red	190
127.	Construction of the thermopile	199
128.	Linear thermopile	199
129.	Galvanometer	200
130.	Heat-curves of the spectra thrown by flint glass and rock salt	201
131.	Action of the invisible thermotic rays	202
132.	Light, heat, and photographic action of the solar spectrum	205
133.	Fresnel's mirror	207
134.	Fresnel's mirror experiment	208
135.	Undulatory ray	216
136.	Interference of two systems of waves	218
137.	Huyghens' principle	230
138.	Explanation of reflexion and refraction	234
139.	Impact of elastic balls	238
140.	Unusual dispersion power of Fuchsin	243
141.	Tuning fork	251
142.	Diffraction or inflection image of a narrow slit	258
143.	Diffraction apparatus	260
144.	Phenomena of diffraction with a circular aperture	260
145.	Phenomena of diffraction with a rhomboidal aperture	260
146.	Explanation of diffraction through a slit	262
147.	Diffraction phenomena through a grating	266
148.	Explanation of diffraction through a grating	267
149.	Comparison of the prismatic with the grating spectrum	271
150.	Newton's colour glass	273
151.	Newton's coloured rings	274
152.	Explanation of the colours of thin laminæ	275
153.	Interference striæ in the spectrum	280
154.	Double refraction in Iceland spar	283
155.	Rhombohedron	284

LIST OF ILLUSTRATIONS. xiii

FIG.		PAGE
156.	Crystalline forms of Iceland spar	285
157.	Double refraction. First case	286
158.	Double refraction. Second case	287
159.	Double refraction. Third case	288
160.	Wave-surface of a negative uniaxial crystal	289
161.	Huyghens' construction of double refraction	290
162.	Two rhombohedra of Iceland spar	293
163.	Polarised ray of light	298
164.	Nicol's prism	304
165.	Polarisation by reflexion	306
166.	Two polarising mirrors	307
167.	Biot's polarising apparatus	307
168.	Nörremberg's polarising apparatus	309
169.	Nörremberg's polarising apparatus, with glass laminæ	311
170.	Tourmaline tongs	314
171.	Parallel Tourmaline plates	315
172.	Crossed Tourmaline plates	315
173.	Two Nicol's prisms employed as a polariser	316
174.	Decomposition of vibrations	317
175.	Dubosq's polarising apparatus	325
176.	Rings of colour produced by uniaxial crystals	328
177.	Rings of colour produced by biaxial crystals	328
178.	Polarisation image of suddenly cooled plate of glass	330
179.	Two Nicol's prisms	332
180.	Rotation of the planes of vibration in Quartz	334
181.	Circular movement of pendulum	336
182.	Decomposition of vibrations	338
183.	Effect of two opposite circular vibrations	342
184.	Double prism of Quartz	344
185.	Tube for the reception of circularly polarising fluids	347
186.	Double plate of right and left rotating Quartz	348
187	Soleil's Saccharimeter	349
188	Compensator	350

OPTICS.

CHAPTER I.

SOURCES OF LIGHT.

1. None of our senses supplies us with such extensive and exact knowledge of the external world as that of sight. The eye penetrates into the unfathomable abysses of space, and receives intelligence from regions the most remote and inaccessible; it reveals to us the delicate cells of which living beings are composed, and perceives the animalcules that people the waters, whilst the manifold forms which it discloses to the mind are rivalled only by the exquisite beauty and charm of colour with which the physical world appears to be decorated.

The visual organ, like every other special sense, possesses a peculiar form of sensibility, that of perceiving luminous rays, a faculty which admits of no more precise definition and explanation than the corresponding sensations of sound or heat, of taste or smell.

The sensation of light can only be excited in our minds by a stimulus of one kind or another acting upon the retina, which is the delicate expansion of the optic nerve lining the posterior part of the eye-ball. The stimulus exciting the sensation may be either mechanical, as by a blow, or by pressure made upon the eye; or electrical, as by the passing of a current of electricity; or it may even be produced by the motion of the blood in the vessels of the retina itself.

External objects can therefore only be perceived by our eyes, or be *seen* by us as the result of something proceeding from them, which reaches our retina, and stimulates it to activity. This something we call *light*.

The science of light (optics) has a twofold problem to solve. On the one hand it has to investigate the laws of light, and on the other to enquire into the phenomena of vision. The former constitutes *Physical Optics*; the latter, *Physiological Optics*. The former, or *physical optics*, is the proper subject of the present course of lectures.

2. Every form of matter when sufficiently heated has the power of emitting rays of light, and thus becomes *self-luminous*. This condition is termed *incandescence*, and the self-luminous worlds, as the sun and fixed stars, are doubtless in a condition of intense incandescence. All *artificial sources of light* depend upon the development of light during incandescence. For the illumination of our streets and houses at night we make use of a combustible gaseous combination of carbon and hydrogen, which forms the chief constituent of ordinary coal-gas. When this hydrocarbon burns, that is to say, when its elements unite with the oxygen of the air, it undergoes, with coincident evolution of heat, partial decomposition. Carbon is separated in the solid state, and floats in a finely-divided and incandescent state in the interior of the burning vapour, and this constitutes the flame. The presence of these

particles of carbon may be easily shown by holding any non-combustible body in the flame, when the carbon in fine powder will be deposited upon it, forming a layer of soot. The combustion of the particles of carbon takes place at the border of the flame, where they are first brought into contact with the oxygen of the air; but if the supply of oxygen to them be insufficient in quantity, they escape in a partially unburnt condition in the form of a dark cloud; and the flame is said to *smoke*.

The brightness of the flame is owing to these solid incandescent particles, for the burning gas itself pos-

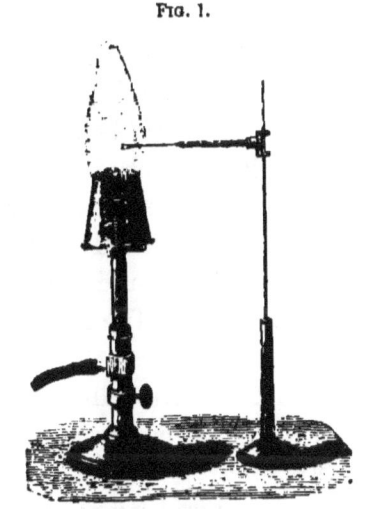

Fig. 1.

Bunsen's burner.

sesses only a feeble illuminating power. This fact may easily be demonstrated by means of a *Bunsen's burner* (fig. 1). In this form of burner ordinary gas conducted through india-rubber tubing streams into the tube of the burner. Air enters, however, through an opening (shown in the adjoining sketch), as well as through a second opening opposite to it, and mixes itself with the gas in the interior of the tube. If the mixture issuing from the tube be now ignited, it burns with an extremely feeble flame which deposits no soot on bodies held in it. For now oxygen is admitted not only to the border of the flame, but throughout its whole mass, and the carbon is accordingly burnt into carbonic acid before it can separate in the solid form, so that the flame is composed of incandescent gases alone. Its illuminating power is therefore very feeble; on the other hand, in consequence of the more perfect combustion that takes place it possesses a far higher temperature than the flame of ordinary gas. It is used as a heat-producing flame, and its temperature can be still further raised by a short conical chimney supported on six metal arms arranged in the form of a star. If a *solid* body be introduced into this feebly-luminous flame, such, for instance, as a piece of platinum wire (see the figure), the incandescent metal glows with a brilliant light. The luminosity of a Bunsen's burner can be restored by shutting off the entry of air, either by closing the holes with the finger or by the rotation of a slide which covers them. The light then becomes much more brilliant, with abundant formation of smoke, its temperature at the same time falling considerably.

The flames of candles and lamps, whether the substance burnt be tallow or wax, rape-oil or petroleum, do not differ essentially from that of an ordinary gas burner. The same hydrocarbon gas which constitutes the essential constituent of common gas is burnt also in them. The hot wick which draws up the fluid material about to be burnt plays the part of a small gas factory, the produce of which is used on the spot. The flames of candles and of lamps all owe their luminosity to the incandescence of particles of carbon floating in them.

3. A petroleum lamp burns, in the first instance, with a dull murky flame, giving off a large quantity of smoke, but it acquires a high degree of luminosity when the glass chimney is applied, for the presence of the chimney causes a strong draught, supplying the air requisite for the thorough combustion of the gas with which it was previously insufficiently intermingled. The brilliancy of a petroleum flame is thus materially exalted by an increased supply of air, whilst that of a Bunsen's burner, as has just been seen, is almost abolished by the same means. The contrary effects observed in these two cases admit of easy explanation. In the latter instance the amount of air supplied is so great that scarcely any of that separation of the particles of carbon takes place, which is so necessary in order that a bright light should be produced. But in a petroleum lamp, the introduction of a moderate quantity of air, by effecting the combustion of the superfluous particles of carbon, causes a higher degree of heat, and consequently a more lively incandescence and illumination of the still remaining particles.

From all this it is obvious that in order to obtain the highest illuminating power of a flame in which hydrocarbonaceous compounds are undergoing combustion, the regulation of the supply of air is essential. A still greater degree of illumination may be obtained, if, instead of air, which only contains one-fifth of oxygen, an appropriate quantity of pure oxygen is conducted into the flame. A burner constructed with this object in view is here shown (fig. 2,) and is termed the *oxygen lamp* or *burner*. In this burner coal-gas flows through the upper horizontal tube into a wide one closed below. Through the middle of this runs a second narrow tube, which is a continuation of the lower horizontal one, and conducts oxygen from an adjoining gasometer. At the orifice the interspace between the two tubes is closed by a funnel-shaped plug, perforated by a series of small openings from which the coal-gas escapes. When this is ignited the oxygen is turned on and enters the interior of the flame, the proportion of the two gases being regulated by means of two stopcocks, shown in the figure. The circular flame can thus be easily rendered intolerably bright.

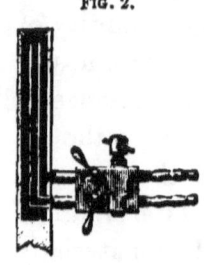

Oxygen lamp.

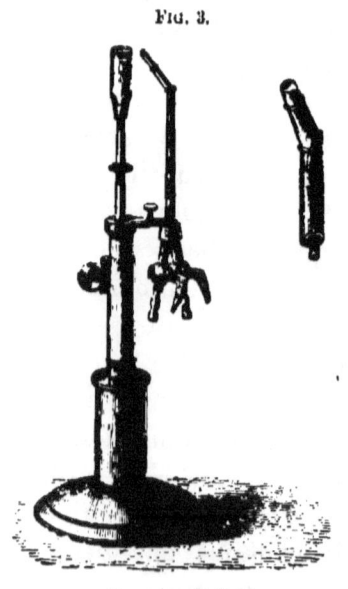

Drummond's light.

4. If more oxygen be admitted than is necessary to produce the greatest degree of illumination, the brilliancy of the flame is diminished, but its heat becomes correspondingly increased in intensity. If a bundle of iron wire be held in the flame the metal burns with vivacity, giving off beautiful sparks and falling in molten drops. On the other hand, if an infusible and incombustible substance, as chalk or magnesia, be introduced into the hot flame, it is raised to white heat and emits

a blinding glare. To obtain this—*Drummond's lime light*, as it has been named, after its inventor—the arrangement (shown in fig. 3), may be conveniently used. Its construction is easily intelligible from what has been previously stated. The bent burner, shown separately at the side, consists of a tube traversed by a smaller one, which last conducts oxygen into the flame of coal-gas issuing from the annular intervening space between the two tubes. The obliquely directed flame plays against a cylinder of magnesia or lime, supported on a convenient stand, and raises it to a white heat. The stop-cocks serve to regulate the proportion of the gases.

5. In the sources of light that have hitherto been considered there has always been a flame; that is to say, a stream of burning gas, by the heat of which a solid body is brought to incandescence and is the cause of the light. In the *Magnesium Lamp*, of which a description will now be given, a *solid body*, magnesium, with its silvery lustre, is burnt in the open air, and the solid product of its combustion, magnesium oxide (magnesia), shines with a splendid light.

The construction of the magnesium lamp made by Salomon and Grant, of London, is represented in fig. 4. A cylindrical box, G, contains two caoutchouc rollers, which, by means of clockwork set in motion by the key c, cause a coil of magnesium wire, on the wheel K, to be slowly unwound and passed through the tube Rf, in proportion to the rapidity with which it is burnt at f. After the end of the magnesium wire has been ignited, the clockwork is set in motion by pressure on the lever m, whilst it is stopped by removing the pressure.

6. None of these means of illumination, however brilliant are those of the lime light and of the mag-

nesium lamp, are comparable with the dazzling light of the electric current passing between carbon points, which is only surpassed by the light of the sun itself.

FIG. 4.

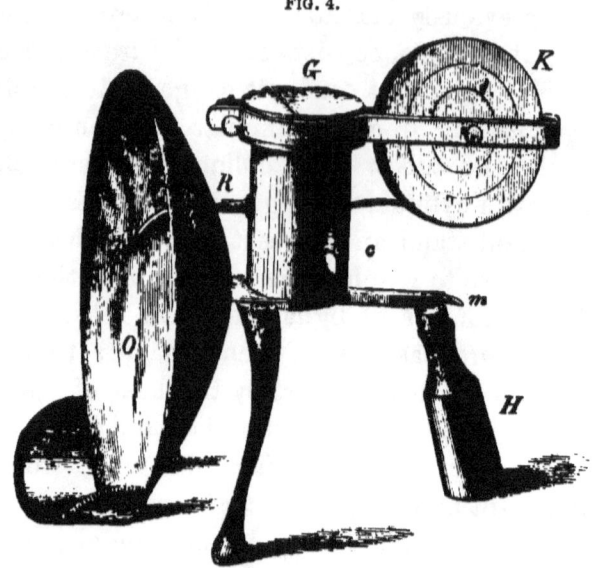

The magnesium lamp.

The apparatus shown in fig. 5 may be used for the production of the electric light. Two metal rods, to the extremities of which pieces of hard gas coke are attached,

FIG. 5.

Electric light between carbon points.

are made to slide through tubes supported on insulating glass stands. Each rod is connected by a wire with one pole of a voltaic battery of about 50 Bunsen's cells. If the carbon points are brought into apposition they become intensely incandescent at the points of contact, and they can then be withdrawn for some distance from each other without interrupting the current or the light it produces.

Between the carbon points an arc of glowing particles of carbon appears, the so-called *Volta's arc of flame*, which effects the conduction of the current at the point of interruption. This flickering arc of flame is far less bright than the carbon points themselves; the particles of carbon of which it is composed detach themselves from the positive pole, which is the hottest of the two, and fly across to the negative pole. As a result of this, after a short time the positive pole becomes shortened and even excavated, whilst the negative preserves its pointed form. At the same time combustion of both poles takes place to a certain extent, owing to the action of the atmospheric air; and the positive pole, which is exposed to the destructive action of two agents, is more rapidly consumed than the negative. The light-phenomena are as brilliant *in vacuo* as in air; and since the combustion of the carbon is thus avoided, the positive pole, which furnishes the particles of carbon for the arc of flame, alone wastes away. This experiment shows that the source of white heat is not here the process of combustion, as in the above-mentioned cases, but results from the glow produced by the electrical current.

7. The resistance which the current has to overcome in passing from one carbon point to the other is greater in proportion as the distance between them increases, owing to their burning away. The strength of the current, however, correspondingly diminishes, till it is

no longer capable of forming an incandescent arc between the opposite poles. The current is then interrupted, and the light dies out. Hence if practical use is to be made of the electric carbon light, it is obvious that care must be taken to keep the carbon points always at a proper distance from each other, and for this purpose apparatuses have been invented which automatically approximate the points in proportion as they are burnt away, and these have been named *carbon-light regulators* or *electric lamps*.

The Regulator of Foucault and Dubosq, the construction of which is shown in fig. 6, is a masterpiece of ingenuity and mechanical adaptation. A complete account of this complicated machine would here be out of place. It will be sufficient to say that by means of clockwork the two carbon points are made to approximate to each other, the inferior (positive) pole moving rather faster than the other, in view of the greater rapidity with which it is burnt off. Before the current reaches this it circulates round the coil of an electro-magnet; as long as the carbon points preserve their proper distance from each other the electro-magnet is sufficiently strongly magnetised to fix an iron detent, and thus to check the clockwork. As soon, however, as the distance between the carbon points, in consequence of combustion, becomes greater, the strength of the current diminishes and the electro-magnet is rendered less powerful—the detent is accordingly set free, the clockwork acts, and the carbon points approximate, which again re-establishes the current in its former intensity; the keeper is then again attracted and the clockwork checked anew. By means of the automatic action of the Regulator, not only are the carbon points kept at a constant and equal distance from each other, but the arc of flame can be maintained unbroken for hours together in the same place.

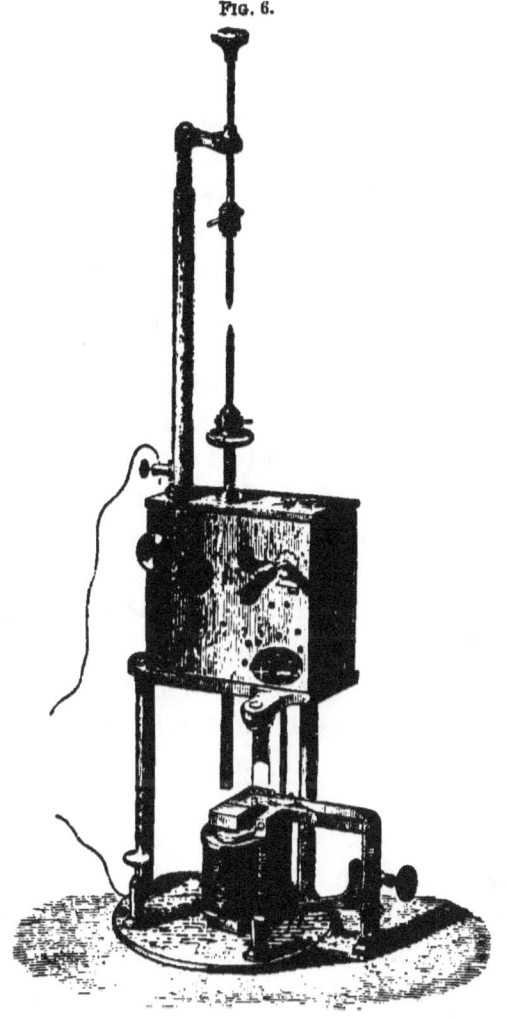

Fig. 6.

Electric lamp.

8. All bodies that do not themselves produce light can only be seen by means of the light they receive and

reflect to our eyes from self-luminous bodies. Amongst the heavenly bodies, the moon and planets are in this case, for they are illuminated by the sun, as are most terrestrial objects. The light falling upon such *non-luminous* bodies is *diffusely* reflected from their surface; that is to say, in such a manner that every illuminated point throws out rays from the surface in every direction.

Every illuminated body, *reflecting light diffusely*, plays the part of a source of light. It shines with borrowed light. Our earth, like the moon and planets, is in this position, in comparison with the self-luminous stars. The faint light which the new moon presents, and which makes that part of her disk visible which is not directly illuminated by the sun, is only the reflection of the earth illuminated by the sun's rays.

9. Light, proceeding from a self-luminous or from an illuminated object, must traverse the humours of the eye before producing a sensation in us by exciting the retina. Bodies which, like the contents of the globe of the eye, or like air, water, glass, etc., permit light to pass through them, are called *transparent*; on the other hand, those which transmit no light are said to be *opaque*. This difference, however sharply expressed it may usually appear to be, is not due to any absolute difference, for every opaque body if reduced to a sufficiently thin film becomes transparent, whilst transparent bodies permit the passage of less light in proportion to their thickness. In the abyss of the sea the darkness of night prevails, because only a sparing amount of light is capable of traversing a mile or more of water. On the other hand, the most opaque bodies, like the metals, can be rendered so thin that a subdued

light glimmers through them. Foucault has, in fact, proposed to cover the object-glass of a telescope intended for solar observation with a thin precipitate of silver, in order to protect the eye of the observer from the glare, without loss of definition.

CHAPTER II.

RECTILINEAR PROPAGATION OF LIGHT.

10. AN opaque body is illuminated on that side of its surface only which is turned towards the light, its opposite surface, as well as a space covered by it, *the shadow*, remains dark. The shadow of a body is projected upon a plane surface held in the shadow-space as a similarly formed dark spot, which occupies that part of the plane to which the access of light is prevented by the body throwing the shadow. It may easily be demonstrated that all straight lines conceived to be drawn from any point of the shadow thrown upon the plane to the source of light, strike against the opaque body, and that only those points of the plane receive light which are so placed that straight lines drawn to them from the source of light are not arrested by the shadow-giving body.

From these facts the conclusion may be drawn *that light proceeding from a luminous body whilst traversing a homogeneous medium is propagated in every direction in straight lines, which are called rays of light.* Those rays which we may conceive to be drawn from the luminous point s (fig. 7), to the circumference of the shadow, graze the surface of the body throwing the shadow and collectively form a cone which invests the body like a ring. The line formed by all the points of contact is the limit between the front illuminated and the back unilluminated surface of the body. The shadow which the object throws upon any plane or curved surface is nothing but the section of this cone

FIG. 7.

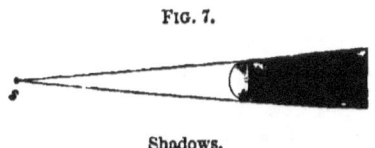

Shadows.

of contact-lines by the plane in question. It consequently holds a direct geometric relation to the form of the object, and forms a simple outline image of it or silhouette. Shadows supply to our eyes, which as it were unconsciously follow the geometric relation between the form of the shadow and that of the object, valuable means for the correct judgment of the real form of bodies in space. The painter uses them to make his figures stand out from the canvas. In technical drawings of machines, scaffolding, etc., which are to serve as plans for the artificer, in addition to the elevation there must always be a 'ground plan,' in order that the perspective relations of the building may be understood. But if in the former the strictly geometric shadows were given, the second might in many cases be dispensed with.

11. If the body casting a shadow be illuminated, not by a single luminous point, as has been supposed in the foregoing illustrations, but by a bright body which possesses innumerable luminous points, we must, in order to know the nature of the shadow, imagine a shadow cone for each luminous point; the space behind the opaque body which is common to all these cones receives no rays from the luminous body and is termed the

nucleus of the shadow; but this is surrounded by a space which is only in shadow as regards a part of the luminous body, whilst it receives light from the rest of it and is consequently partially illuminated. It is termed the *half shadow or penumbra*. Fig. 8 shows the case of a large luminous sphere, *A*, opposite which is a smaller opaque one, *B* ; the simple construction shows what determines

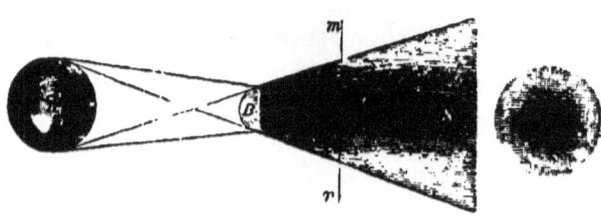

Shadow nucleus, and penumbra.

the limits of the nucleus of the shadow and the penumbra. The conical nucleus of the shadow terminates in a point at *S*, whilst the penumbra stretches away constantly widening to infinity. A plane held in the shadow at *m n*, perpendicular to the axis of the cone, receives the image represented in fig. 9, where a central dark spot is seen corresponding to the nucleus of the shadow, and is surrounded by a less dark area, the shade of which gradually diminishes from within outwards till it is no longer perceptible. If the plane be closely approximated to the body giving the shadow, the broad dark nuclear shadow loses but little of its definition, the half shadow surrounding it appearing only as a narrow border. If placed at a greater distance, the penumbra exceeds the nucleus of the shadow in breadth, and only an ill-defined shadow results. An explanation is thus afforded why we are unable to point out the exact spot where the shadow of a steeple ends on the ground. So if a knitting-needle be held in the sun immediately in front of a sheet of paper, it throws a very well-defined shadow; but if it be removed to a distance of only three or four inches from the sheet no accurate outline can be traced of its ill-defined shadow.

Our planetary system affords striking illustrations of such shadow cones as are shown in fig. 8. The shadow nucleus behind the moon is nearly equal to the radius of the moon's orbit, and can, therefore, when the moon intervenes between the sun and the earth, which is only possible at the time of the new moon, reach the surface of the earth. The sun is then totally covered by the moon, or there is said to be a *total eclipse of the sun* over those parts of the earth which are in the nuclear shadow; whilst in those parts which lie in the penumbra a sickle-shaped portion of the sun's disk remains visible, and the eclipse is only a *partial* one.

The nuclear shadow of the earth extends behind it to a distance of 216 of its semidiameters, and thus reaches far beyond the radius of the moon's orbit, which amounts to only 60 semidiameters of the earth. At the time of the full moon it may happen that the moon lies wholly or partially in the earth's shadow, and the interesting spectacle of a *lunar eclipse* is presented to us.

12. To an observer placed at the point S of the cone (fig. 8), the smaller but nearer sphere B appears to be of exactly the same size as the larger but more remote sphere A, the latter being precisely covered by the former. The *apparent size* of an object is determined by the angle which the rays of light, passing from its outermost points to the eye, form with one another,

the so-called visual angle. The same body is seen under a smaller visual angle, and of correspondingly smaller size the further it is removed from our eyes, and two bodies of different size appear under the same visual angle if their distances are inversely as their diameter. If we are acquainted with the real size of an object we can determine its distance from us by the visual angle under which it appears to us; and, *vice versâ*, if the distance and the apparent size be given, we can determine its actual size. Astronomers employ these simple data to determine the size and distance of the heavenly bodies. It has been found, for example, by appropriate observations, that the semidiameter of the earth, seen from the sun, would appear under a visual angle of only 8' 6". This is termed the *parallax* of the sun; and from thence the calculation has been made that the distance of the earth from the sun amounts to 24,000 semidiameters of the earth, and after this distance is determined it results, from the visual angle of 32' under which the sun appears to us, that its diameter is 112 times greater than that of the earth.

The same operations by which the astronomer obtains his results school us from our youth upwards to form every day and every hour an unconscious estimate of the size and distance of terrestrial objects by the *measurement of the eye*. The visual angle under which a human form or other object of known size appears to us supplies us with a datum from which we estimate its distance, and this distance again enables us to form a judgment in respect to the size of neighbouring objects. The rays of light which reach the microscopically small earth from the various parts of the mighty mass of the sun, do not form a greater angle with each other at most than 32', which expresses the apparent size of the sun, and may therefore be regarded as being almost parallel. If a beam of the sun's rays be allowed to enter a chamber through a wide opening in the window shutter, it may be easily followed by the illumination of the floating particles of dust, and it may be shown that it has everywhere the same diameter, and must consequently be composed of parallel rays.

13. If now the chamber be completely darkened, and a very small opening of from 1–3 millimètres ($\frac{1}{25}$th–$\frac{1}{8}$th of an inch) be made in the shutter, a very pretty appearance may be observed upon a paper screen placed opposite to the opening. The neighbouring buildings are seen with their roofs, chimnies, and windows; the green tree tops waving in the wind, men walking in the streets, white clouds sailing over the blue sky, in fact a complete picture of the external world is as it were painted in delicate colours upon the screen. But this picture is inverted; what is in reality above appears in the picture below, what is there on the left is here on the right, and *vice versâ*. When the screen is brought nearer to the opening, the picture becomes smaller but clearer; when it is removed to a greater distance it becomes fainter but its size is increased. If the circular opening be replaced by a square one of equal area, the picture undergoes no change, nor does any alteration occur if the square be changed to a triangle of equal area; but when, on the other hand, a series of continuously larger and larger openings be used, the picture will be found to become progressively brighter, whilst its outline becomes more and more confused and blurred, until, when the opening is several centimètres

in diameter, no definite picture can be discerned upon the screen, but only a uniformly illuminated surface.

The mode of production of this charming picture is best explained by a repetition of the same experiment in a simpler form. A lighted candle is placed in front of a screen perforated by a small opening (*O*, fig. 10), and behind it a white paper screen (*S*) is held which receives the inverted image of the flame. Amongst the

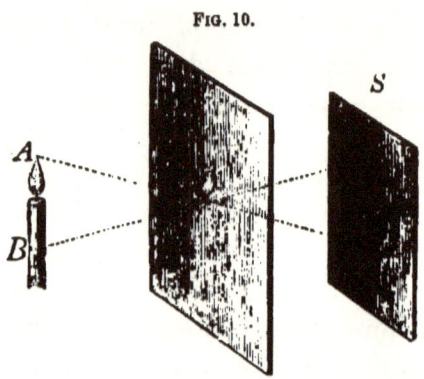

FIG. 10.

Projection of an image through a small aperture.

innumerable rays of light which, for example, the highest point, *A*, of the flame emits, only a small conical fasciculus (*A a*) traverses the aperture and forms upon the screen a small bright spot (*a*) which, in consequence *of the rectilinear course of the rays of light* is only illuminated with the light of the point *A*, whilst no other part of the screen can receive light from this point. In the same way, the spot *b*, situated upon a higher part of the screen, is only illuminated by the lower point, *B*, of the object. Now since every point of the object sends its luminous rays separately to different points of the screen, the continuous serial addition of innumerable bright spots forms an image which, as is immediately intelligible from the figure, resembles the object, and is larger in proportion as the screen is removed from the aperture. The larger the image, the feebler is its illumination, because the same quantity of light is then distributed over a larger surface.

The small spot of light, a, must necessarily be circular or square or triangular, in accordance with the shape of the opening. But since the adjoining light spots overlap each other, its particular form is of no importance; and the result is the same in regard to the entire image, whatever may be the form of the aperture. If the rays of the sun penetrate through a partially closed window shutter they throw upon the floor of the room bright elongated and rounded spots of light. These are so many images of the sun's disk thrown by the various irregularly formed chinks and apertures of the shutter. The illuminated spots do not appear circular but elliptical, because the surface of the floor on which they fall is not perpendicular to the direction of the sun's rays. The spaces between the leaves of the thick foliage of a tree act in the same way, and produce numerous elliptical images of the sun on the shaded floor of the forest. In partial eclipse of the sun these light-spots in the shadow thrown by trees assume a distinctly sickle-shaped form.

It is now obvious why *small* openings are alone capable of forming such images, for they only are capable of effecting such a division of the rays of light as is essential for the production of an image: large openings, which allow rays of light to fall upon the screen from all or very many points of the object, are not appropriate for the purpose.

14. If there be a luminous point at L (fig. 11), and a, b, c, d be an opaque screen, A, B, C, D would be the shadow which this screen would throw on a second screen placed parallel to it. If the second screen be just twice as distant from the source of light as the first, the area of the shadow will be *four times as large* as the screen which throws the shadow. If the latter be removed, the same number of rays, which was previously received by it and illuminated its surface, is now distributed over an area of four times the size; a

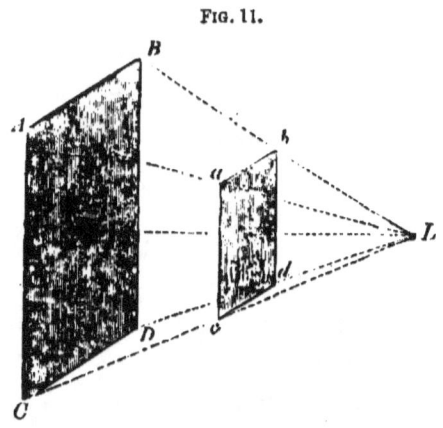

FIG. 11.

Diminution of the illumination in the ratio of the square of the distance.

given portion of the surface A, B, C, D, receives, consequently, four times less light than a corresponding portion of the surface a, b, c, d, and will be therefore proportionately less strongly illuminated. The source of light thus gives, at double the distance, only the fourth part of the illumination which it can give at unity. If the second screen be at 3, 4, 5 . . . times the distance of the first from the source of light, the shadow falling upon it will be 9, 16, 25 . . . times larger than the shadow-throwing screen, and will, according to its distance, be 9, 16, 25 . . . times less brilliantly illuminated.

We thus acquire a knowledge of the law, that *the amount of illumination diminishes in proportion to the square of the distance from the source of illumination.*

The apparatus shown in fig. 12 may be employed to demonstrate the truth of this law by experiment. A sheet of white paper is stretched on a frame, supported on a stand S, in the centre of which is a spot of oil, made with stearine. The grease spot allows more light to pass through it, and consequently reflects less

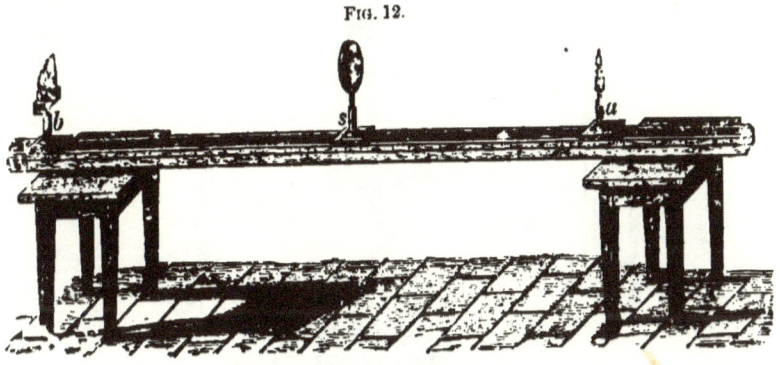

Fig. 12.

Bunsen's Photometer.

than the unstained part of the paper. If therefore the paper be illuminated more strongly from behind, it appears bright on a dark ground. On the other hand, it appears dark upon a bright ground if it be more strongly illuminated on the front surface; whilst, with equal illumination on both sides, the spot becomes invisible, since it can then appear neither darker nor lighter than the adjoining paper. The flame of a candle, a, is now placed upon one side of the screen, whilst four such flames are placed upon the other side at b, and the screen is removed to such a distance from them that the spot is no longer visible. This will be found to occur when the distance of the quadruple flame from the screen on the one side is double that of the single flame on the other side. This experiment, in which a source of light four times as strong as another gives the same illumination at double the distance, corroborates the law above laid down.

This law being admitted, the same apparatus, fig. 12, may be employed as a means of comparing the brilliancy of two sources of light. If, for example, the flame of a candle be placed in front and a gas flame behind a paper screen, and this be moved till the grease

FIG. 13.

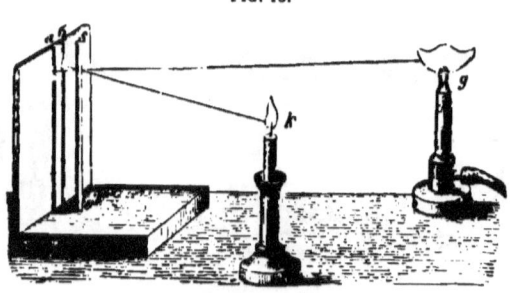

Rumford's Photometer.

spot disappears, the illuminating power of the two lights will be as the squares of their distances from the screen. The apparatus employed for the determination of the illuminating powers of different sources of light, are termed Photometers. The paper screen with the grease spot constitutes the essential feature of the Photometer of Bunsen.

Rumford's Photometer is of remarkably simple construction (fig. 13). An opaque rod, about the size of a lead pencil, stands in front of a white paper screen.

The two lights to be compared both cause a shadow of the pencil, and each light illuminates the shadow cast by the other. If either light is removed to such a distance that the two shadows appear of equal depth, the brilliancy of the two lights will be as the squares of their distances from the screen.

CHAPTER III.

REFLEXION OF LIGHT.

15. If a beam of parallel rays of light from the sun be allowed to pass obliquely through an opening in the window shutter ($f\,n$, fig. 14) and to fall upon the plane surface of mercury at rest ($s\,s'$), it will be seen that from the point (n) where the beam strikes the surface of the mercury, a second fasciculus of rays ($n\,d$) proceeds, the course of which may be followed just as easily as that of the incident ray, by its illuminating the floating particles in the air.

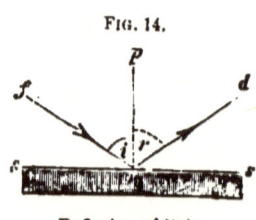

Fig. 14.

Reflexion of light.

This process is termed *regular reflexion*, in opposition to *diffuse reflexion*, which has been already referred to (p. 12). If a sheet of paper be placed upon the mercury, the reflected beam vanishes, but the spot, n, where the paper is struck by the incident rays is brilliantly illuminated and becomes visible from every side as though it were self-luminous. The *dull* surface of the paper, although it may be struck in a certain direction only by rays of light, thus emits rays in all directions, and becomes in virtue of this *diffuse reflexion* everywhere visible as an illuminated object. The *smooth* surface of the mercury, on the other hand, appears not at all or but very feebly illuminated at the point *n* where it is struck by the incident rays; it *reflects* them in a *perfectly definite direction* without otherwise materially altering them. In fact, if a sufficiently small opening be made in the shutter, the same oval image of the sun appears on the roof of the room where the reflected ray falls, as the incident ray itself would have formed had it been allowed to fall upon the floor.

Every smooth surface is called a *mirror*, and Nature herself offers to us, in the surface of fluids at rest, a very perfect example of a mirror. Mirrors, however, that are composed of some solid material, as of polished metal, although this can never be made to attain the absolute smoothness of the surface of a fluid, are very much more convenient for use. The kind of mirror most commonly employed consists of a plate of glass which has been ground and polished and covered on one surface with an amalgam of tin, or with a precipitate of silver, and the surface of the metal adhering to the glass is generally the reflecting surface.

In order to indicate accurately the course of the incident and reflected rays, we must conceive a vertical line, or *perpendicular* (*n p*), to fall on the reflecting surface at the point *n* (fig. 14) where it is struck by the incident ray. The plane drawn through the incident ray and the perpendicular, which is itself vertical to the plane of the mirror, is called the *plane of incidence;* it is also named the *plane of reflexion, because it always contains the reflected ray.* The path pursued by the incident and the reflected rays is determined by the *angle of incidence*, *i*, and the *angle of reflexion*, *r*, which each of the rays make with the perpendicular. *The angle of reflexion is always equal to the angle of incidence.*

These two propositions—that the planes of incidence and reflexion are coincident, and that the angles of incidence and reflexion are equal—together constitute the no less simple than important *law of the reflexion of light*. In order to demonstrate it by experiment, the instrument shown in fig. 15 may be used. To the curved border of a semicircular piece of wood, *A A*, a plate of metal is attached which has a vertical slit at the centre of its curve (*a*), and from this point outward is divided into 90°. The mirror *f*, the back of which is shown in the figure, is capable of being rotated round a vertical axis,

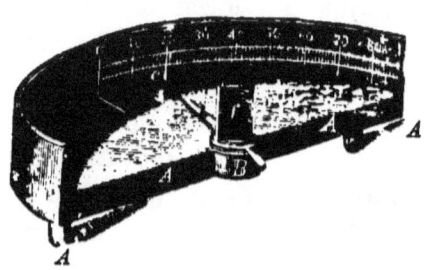

Model for the demonstration of the law of reflexion of light.

B, passing through the centre of the semicircle. The rod *b*, which is attached to the mirror and points by means of an indicator, *c*, to the scale of degrees, is at right angles to the plane of the mirror, and consequently represents the perpendicular. If now a small beam of parallel rays be allowed to pass through the slit and fall on the mirror, the reflexion will illuminate and make visible that part of the circumference of the circle towards which it is directed. The indicator *c* now stands, we will say, at 20°. The ray coursing from *a* to *f* strikes the mirror under an angle of incidence of 20°, and hence if the above law of reflexion be correct, should be reflected to the line marking 40°, and in point of fact it will be found that this is the degree which is brilliantly illuminated by the reflected light. If now the indicator be successively placed opposite the lines marking 10°, 20°, 30°, etc., the reflected ray will successively illuminate the lines marking 20°, 40°, 60°, etc., as the law of reflexion requires that it should do. If, lastly, the indicator be placed opposite the slit itself, so that the angle of incidence is zero, the angle of reflexion must also be zero; the reflected ray passes out again by the slit in the same direction as the incident ray entered, or in other words, *a ray of light falling perpendicularly upon a mirror is reflected upon itself.*

16. A *plane mirror* reflects the images of objects

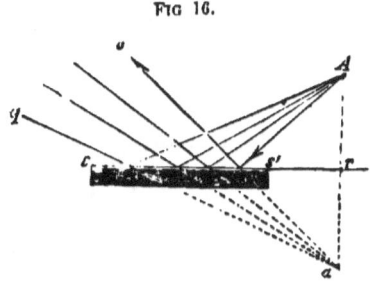

FIG 16.

Production of the image point in a plane mirror.

placed in front of it, ourselves included, with an accuracy that is proverbial. The production of these images may be explained in the simplest manner by the law of reflexion. In the diagram (fig. 16) An and Ap represent two out of the innumerable rays which a luminous point A throws upon a mirror ss'. If we conceive the reflected rays, $n\ o, p, q$, corresponding to them, and the direction of which, in accordance with the above law admits of being easily ascertained, to be prolonged backwards, they will meet each other in the point a. The straight line $A\,a$, which joins the point a with the luminous point A, is perpendicular to the plane of the mirror and is bisected by it at the point r, that is to say, $a\,r = A\,r$, which is deducible also from the fact that the triangles $A\,n\,r$ and $a\,n\,r$ are equal to one another. Since any pair of rays, that may have been selected for consideration, pass to the same point, a, it follows that all the rays proceeding from A that fall upon the mirror can similarly be carried back as though they proceeded from the single point a. We can therefore make the following proposition as a direct corollary of the law of reflexion:—

All rays that proceed from a luminous point and fall upon a plane mirror, are reflected from it as if they came from a point in a perpendicular dropped from the luminous point to the mirror, as far behind the reflecting surface as this is in front of it.

An observer placed in front of the mirror receives consequently the reflected rays as if the point a, from which they appear to proceed, were itself the luminous point. It sees in, that is to say, behind the mirror, the point a as the image of the luminous point A, situated in front of the mirror.

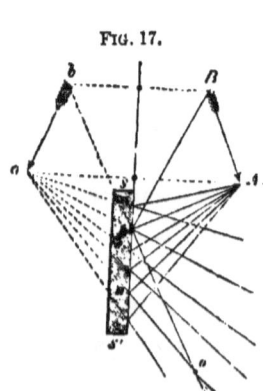

Fig. 17.

Production of the image in a mirror.

In the same way an image point behind the mirror corresponds to each point of every luminous or illuminated object, and out of the totality of the image-points the complete mirror image or reflexion of the object is produced. In order to conceive this image

REFLEXION OF LIGHT. 31

in the mind, or to show it in a drawing (fig 17), a perpendicular must be conceived to be struck from each point of the object to the plane of the mirror, and prolonged as far behind it as these points are in front of it. In such a simple object as an arrow, AB (fig. 17), which may be selected as an example, it is only requisite to show the construction for its terminal points, A and B, by which its image ab is formed. An observer situated at o receives the rays from the point of the arrow in the direction $A\,n\,o$ and from the other extremity in the direction $B\,p\,o$. Simple inspection of the figure shows that the image and the object must be of equal size, and must also lie symmetrically with regard to the plane of the mirror.

17. The polished surface of this plate of glass (fig. 18)

FIG. 18.

Mirror-image in a transparent plate of glass.

acts as a mirror, whilst at the same time it permits the objects behind it to be seen. If a lighted candle be placed on one side the image is reflected. If a water carafe filled with water be placed behind the glass plate in the apparent position of the image, the illusory impression is produced of a candle burning whilst submerged in the interior of the flask. In this simple experiment lies the explanation of the recently attractive 'Ghost phenomena.' In this class of illusions the back part of the stage is closed by means of a very large transparent piece of plate-glass, somewhat inclined forwards, through which the audience perceive the players feebly illuminated. The 'ghosts' with which they appear to communicate are the reflected images of other persons who are concealed from view, and are in front of and below the stage; these, however, in order to give sufficiently bright reflected images, must be illuminated by the electric or lime light.

18. In order to direct the rays of the sun into the room in a convenient, that is to say, in a horizontal

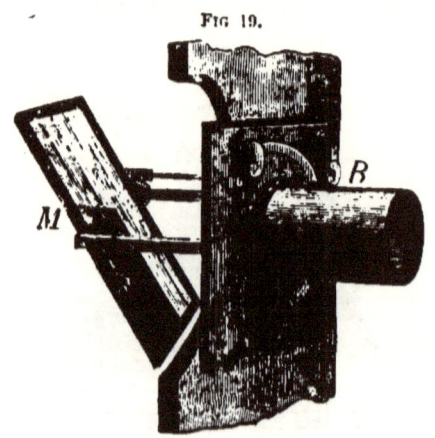

Fig 19.

A Heliostat.

direction, a plane mirror is employed. To the opening in the shutter is attached a board (fig. 19) on the inner side of which is a wide horizontal tube, containing the apparatus intended to be used; externally is a mirror, M, which can be turned on an axis passing between two rods. The mirror can be rotated on the one hand around the axis of the tube by moving the button *A* in a semicircular slit, and on the other hand it can be inclined to the tube at any angle that may be desired by turning the button *B*, which acts on the previously mentioned axis of the mirror by means of an endless screw and rack. It is an easy matter to direct the reflected rays of the sun through the tube by manipulating the buttons *A* and *B*, and to maintain them in that direction notwithstanding the progressive movement of the sun. This apparatus is termed a *Heliostat*.

The perpetual correction of the position of the mirror by means of the hand is, however, not only troublesome but far too uncertain and unsatisfactory for all experiments requiring great steadiness in the direction of the incident rays. A Heliostat has accordingly been constructed, the mirror of which is constantly presented to the sun in the same position by means of clockwork. Fig. 20 shows the Heliostat of Reusch. The axis of the clockwork on which the lower mirror is supported is placed parallel to the axis of the earth, around which, during the daily revolution of the earth, the vault of heaven, and with it the sun, appears to turn. The mirror is then so placed that the reflected

Fig. 20.

Heliostat of Reusch.

rays of the sun course in this axis, and are kept unaltered in it by the movement of the clockwork. By means of a second mirror placed above, capable of being moved into any position that may be required, the beams of light can be made to travel in the desired horizontal direction.

19. The principle of the method, based on the reflexion of light, by which the angles of the surfaces of prisms, crystals, etc., are measured may now be

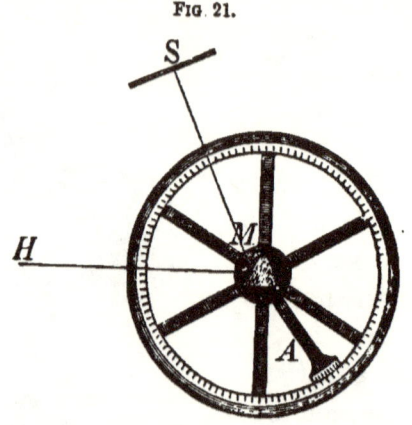

Fig. 21.

Principle of the Reflecting Goniometer.

described. Fig. 21 represents a horizontal circle, divided at its border into 360°; at its middle is a small plate, M, which revolves, and with which an indicator, (Alhidade) A, pointing to the divisions, is connected. A glass prism is placed upon the plate M in such a position that its angles and polished surfaces are vertical. A small beam of the parallel rays of the sun, directed into the chamber through a vertical slit by means of a Heliostat, is reflected from the anterior surface and forms a bright vertical line upon a screen, S, placed at the side. The indicator, A, and with it the prism, is now turned until a second surface of the prism reflects the rays in the same direction, that is to say, until the bright line occupies the same position on the screen. The second surface must now of course occupy the same position as the first was in previously. If the second surface be parallel to the first, it is obvious that the indicator must revolve through 180° to bring the bright spot to the same place, but if the second surface forms with the second any angle a, the object is attained by a revolution of $180-a$ degrees. In order therefore to obtain a knowledge of the angle a between the two surfaces of the prism, it is only necessary to subtract the angle of revolution of the indicator, which can be read off on the divisions of the circumference, from 180°.

Instruments constructed on this principle, and adapted for the exact measurement of the angles at which the surfaces of prisms are placed to one another, are called *reflecting goniometers.*

20. As the reflected rays proceed from the image behind a mirror exactly as they would from an object placed in that position, every reflected image must act as a material object in regard to a second mirror, and this again is in a position to furnish a reflected image. By arranging *two* mirrors so that their reflecting surfaces are turned towards each other, there are produced, besides the two reflected images of the first order, still others of the second, third, and higher orders, which, however, continually become fainter in consequence of the loss of light. Hence when a lighted candle is held between two mirrors placed opposite to one another, we see an indefinite succession of flames which appear to be lost in infinite distance. The number of reflections becomes limited as soon as the two

mirrors form an angle with each other. In fig. 22 the two mirrors furnish the reflexions of the first order, *B* and *B'*, of the object situated between them. Since the image *B* behind the first mirror sends its rays to the second mirror, this gives an image or reflexion of the second order, *C'*, and similarly, the first mirror gives a reflexion, *C*, of the image *B'*. An observer (*O*) placed between the mirrors sees the reflexions, in addition to the object, regularly disposed upon a circle described around the point of decussation of the two mirrors, an image appearing at each angle space which is equal to the angle of the two mirrors. The observer, *O*, therefore, sees the object as often as the angle between the two mirrors is contained in 360°.

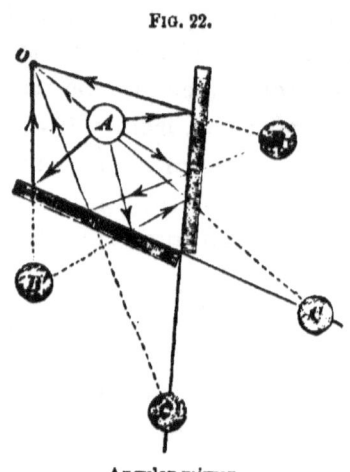

FIG. 22.

Angular mirror.

The pretty effects obtained in the well-known plaything termed the *kaleidoscope* result from the regular disposition of the images reflected by mirrors placed at an angle. An instrument of this kind may be purchased for a few pence in every toyshop. It is composed of a papier-maché tube in which are two mirrors inclined to one another at an angle of 60°. To the front end is attached a cap, capable of being rotated and containing in its interior two plates of glass, the outer one of which is ground dull. Between the two plates are a number of pieces of differently coloured glass, and other small variegated objects. If the tube be placed in a horizontal position, and the plate of ground glass be illuminated with a powerful light, a six-rayed star will be seen upon the opposite screen, decorated with the richest ornamentation.* This is the reflexion in the mirror of the fragments of glass which are combined to form this regular mosaic. If the cap be turned, the pieces of glass constantly form new combinations, and thus an inexhaustible succession of the most delicate forms are obtained which the liveliest fancy could scarcely invent. What may in this way be represented for a large number of persons, as if it were an object on the screen, can also of course be seen separately by every one who looks into the tube for himself.

21. Not only this ingenious plaything, but an in-

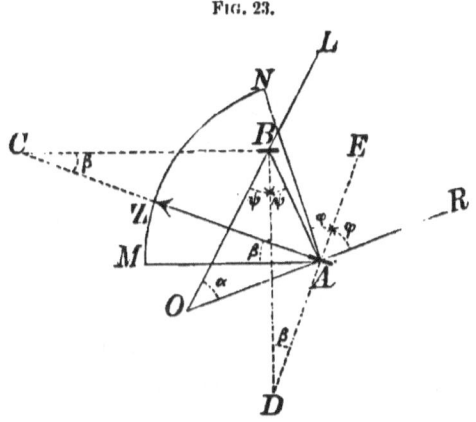

Fig. 23.

Principle of the mirror sextant.

strument of high practical value, is founded on the mutual action of two mirrors placed at an angle to one another. In fig. 23, A and B are two small mirrors,

* In this experiment a lens of short focus is placed at the front end of the kaleidoscope.

the reflecting surfaces of which are turned towards each other. If two objects are placed at L and R, of which the former is visible to an observer at O, above the edge of the mirror B, in the direction OB, the mirror A may have such a position given to it that the light coming from R reaches the eye after double reflexion in the direction $R A B O$, and consequently two objects are seen in the same direction, $O B$, the one direct, the other reflected. Thus it results from the law of reflexion that the angle a which is included by the visual lines extending from the eye to L and R, is *exactly twice as large as the angle β which the two planes of the mirrors form with one another.** In order to measure the angle β conveniently, the mirror A is made to rotate around the axis of a divided arc, $M N$, and is connected with an indicator, $A Z$. The mirror B is permanently fixed on the plane of the arc parallel to the radius $A M$ which goes to the zero of the division. If any object, L, be now looked at in the direction $O L$ through a telescope attached to the instrument (fig. 24) and the indicator, and with it the mirror, be rotated until the image of R is seen in this direction, twice the angle read off by the

Fig. 24.

Mirror or reflecting sextant.

* If the perpendiculars $A E$ and $B D$, which if sufficiently prolonged cut one another in the point D at an angle β are erected upon the mirror planes, it follows if ϕ and ψ indicate the angle of incidence of the rays $R A$ and $A B$ on the mirrors A and B from the consideration of the triangle $A B D$

$$\beta = \phi - \psi$$

and from the consideration of the triangle $A O B$

$$a = 2\phi - 2\psi$$

from whence it immediately follows that $a = 2\beta$.

indicator immediately gives the angle which the visual lines directed towards L and R form with each other.

This ingenious angle measurer, conceived by Newton and constructed by Hadley, is termed the *reflecting sextant*. It is superior to other instruments made for this purpose because it needs no support, but during the act of measuring the angle can be held freely in the hand. Hence for nautical purposes it is the only available angle-measuring instrument. By means of the reflecting sextant the seafarer makes those measurements by which he determines the latitude and longitude of his ship. The two almost invisible mirrors enable him to pursue his predetermined course through the pathless waste of waters.

CHAPTER IV.

SPHERICAL MIRRORS.

22. A spherical shell, the inner surface of which is highly polished, is called a *spherical concave mirror*. It may be regarded as a portion of a hollow sphere cut off by a plane $M M'$, fig. 25. A perpendicular, $c d$, let fall from the centre, c, of the sphere of which the mirror is a segment upon this plane, will strike the middle point of the mirror, and is termed its *principal axis*. The angle $M c M'$, which the lines $M c$ and $M' c$, drawn from two diametrically opposite points of the periphery of the mirror to the centre of the sphere form with one another, is called the *aperture of the mirror*. In practice, only mirrors of small aperture are in use, in which this angle amounts at most to six or eight degrees, and the remarks here made will only have reference to these.

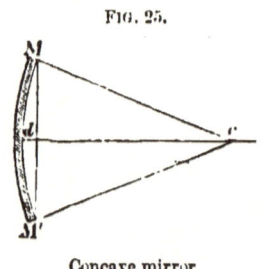

Concave mirror.

If a beam of parallel solar rays, thrown horizontally into the chamber by means of the Heliostat, be allowed to fall upon a concave mirror with small aperture parallel to its axis (fig. 26), it will be seen—for the path of the rays can be distinctly followed by the illumination of the particles of dust always present in the air of a room—that it is reflected in the form of a cone of light, the apex of which, *F*, lies in front of the mirror in this axis. This point, *F*, through which all the rays falling on the mirror parallel to its axis pass after reflexion, is called the *focus*. It becomes brilliantly luminous if I throw some dust into the air in its vicinity. It appears as a white spot of dazzling brilliancy when a white sheet of paper is held in it, and the wreaths of smoke that are now rising from it show you that the paper has caught fire in the intense heat of the rays collected at this point, and that it has consequently been appropriately named the focus or burning-point (Brennpunkt). The space intervening between the focus and the mirror—the focal distance—can easily be measured, and is found to be equal to half the radius of curvature of the mirror, or in other words, the focus lies midway between the mirror and the centre of the circle of which it is a segment.

FIG. 26.

Focus.

23. The reflexion of a ray of light from a curved surface follows the same law as from a plane surface; the portion of the curve which immediately surrounds the minute point of incidence on that each ray of light impinges can alone be considered to act as a reflector. The smaller we admit the superficial area of this part to be—and we may conceive it to be as small as we please—so much the more accurately can we regard it as a small plane mirror, and the perpendicular erected upon this is then the axis of incidence in regard to which the incident and the reflected ray behave as has been

stated. Concave differ only from plane mirrors in the circumstance that each point has its own axis of incidence.

Since every radius of a spherical surface is perpendicular to the surface where it meets it, we obtain the axis of incidence of a spherical concave mirror by simply drawing the corresponding radius to the point of incidence.

In a concave mirror of small aperture *the axes of incidence, that is to say, the radii, are more and more strongly inclined to the principal axis in proportion as the corresponding points of the mirror are more distant from it.* Hence every ray of light running parallel to the axis must be inclined from its original direction more and more strongly towards the axis in proportion as it strikes the mirror at a point more distant from the axis. This, which is clearly exhibited in fig. 26, explains why all rays falling on the mirror parallel to its axis must pass through a single point after reflexion.

24. From the above-mentioned direction of the axes of incidence, it follows further that all rays proceeding from a point pass through a single point after reflexion, because they undergo a change in their direction greater in proportion as the point of the mirror struck is distant from the principal axis.

FIG. 27.

Conjugate foci.

In the concave mirror, fig. 27, which is supported on a stand, two indicators (omitted in the figure) point to the principal focus F, and the centre of the sphere c.

At the point *A* in the axis, the light of an electric lamp is placed, which, to protect the eye from its glare, is enclosed in a box having only a round opening on the side turned towards the mirror. A diverging cone of rays proceeding from the luminous point *A* passes to the mirror and is reflected forwards from it as a converging cone, the apex of which lies at *a* in the axis of the mirror, between the focus and the centre of the sphere. This point of union of the reflected rays is called the image of the point *A*. If the luminous point *A* be approximated to the mirror, the point at which the rays unite, *a*, retreats from the mirror towards the centre *C*; if the luminous point be placed at *C*, every ray it emits strikes perpendicularly upon the surface of the mirror, and is therefore reflected upon itself; and thus, when it is situated in the centre of the circle of curvature, the light and the reflected image of the light are coincident. If, on the other hand, the light be removed from *A* to a greater distance from the mirror, its image continues to approach the focal point, and would ultimately coincide with it were it possible to remove the light to an infinite distance. The removal of the luminous point to infinite distance, which it is of course impossible to accomplish, has been effected, however, in the foregoing experiment (fig. 26), for rays which run parallel with the axis may be regarded as coming from a point on the axis at an infinite distance, and they are, as has been seen, united in the focus.

It is further intelligible that rays of light which, proceeding from the point *a*, strike upon the mirror, are reflected to the point *A*, pursuing the same course but in the opposite direction; in other words, if a luminous point *a* lies between the focus and the centre of the sphere, its image is situated at A on the other side of the centre. The two points, A and a, are thus so associated that *each constitutes the image of the other*, and they are hence called corresponding or conjugate points. To the focus itself consequently, an infinitely remote point is conjugate; that is to say, rays which proceed from the focus and strike the mirror are reflected parallel to the principal axis to an infinitely remote distance. If we place the luminous point (A, fig. 28) nearer than the focus to the mirror, this is no longer capable of collecting the

FIG. 28.

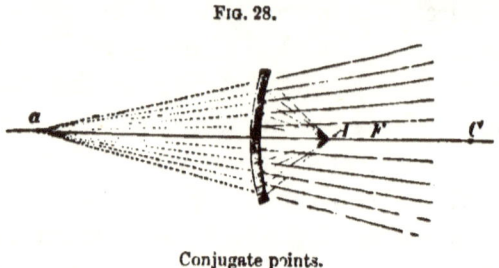

Conjugate points.

too strongly diverging rays, and the reflected rays diverge as if they proceeded from a point a, situated behind the mirror; and so conversely, since rays which converge towards a point a behind the mirror, are united in the point A in front of the mirror, the two points A and a may be regarded as conjugate points.

25. Hitherto the case of luminous points lying in the principal axis of the mirror has alone been considered. The electric lamp must now be placed in such a position that its luminous point lies above the axis (at A, fig. 29). It will then be seen that the reflected rays unite in a single point B, but this lies below the axis on the straight line which may be conceived to be drawn from the luminous point A, through the centre of the

sphere C, to the mirror. Amongst all the rays which proceed from A and strike upon the mirror, that passing through C is the only one that falls perpendicularly upon the mirror, and is therefore reflected upon itself. The straight line, AC, holds therefore the same relation to

FIG. 29.

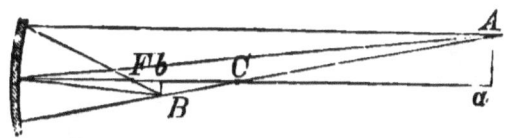

Conjugate points on a secondary axis.

the principal laterally-situated point A as the axis, CF, has to the previously-considered position of the luminous point; it is termed therefore the *secondary axis* corresponding to the point A. For every secondary axis, the number of which is of course infinite, the same holds that has already been stated in reference to the chief axis, each, for example, has its own focus in which the rays parallel with it meet.

The peculiarities of concave mirrors, as far as they have hitherto been considered, may be summed up in the following propositions: All rays that, before they fall upon the mirror, proceed from a point or travel towards a point, pass, after reflexion, through a single point (either actually or when prolonged) which lies on the axis corresponding to the first point. These two points are so conjugated that the one is the image of the other.

26. Inasmuch as to every point of a luminous or illuminated object situated in front of a concave mirror there is a corresponding image-point situated on the axis belonging to it, it follows that from the collection of all the image-points an image of the object results.

46 OPTICS.

Now let a lighted candle be placed *between the focus and the centre of curvature* of the mirror* (fig. 30). The place of the image can easily be found by moving to and fro a paper screen, situated on the other side of the centre of curvature, and protected from the direct rays

FIG. 30.

Real image.

of the flame by a small blackened metal disk. An *inverted and enlarged image* of the flame is then obtained upon the screen, as is shown in fig. 31, in which the course of the rays of light for the point B of the object $A B$ is indicated, showing how the inverted enlarged image $a b$ is formed.

FIG. 31.

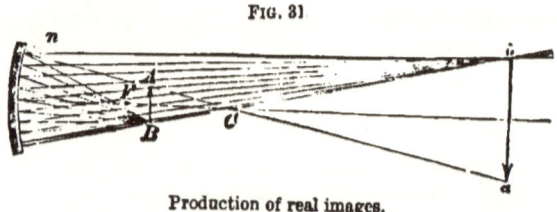

Production of real images.

If, as in this figure, all the points of the object are found in a single plane ($A B$) perpendicular to the axis, the points of the image (always presupposing the

* In the figure the focus is found over the number 132, the centre of the curvature over 120.

aperture of the mirror to be small) lie also in a plane perpendicular to the axis. It is obvious also, from the drawing, that image and object are similar to each other, and their relative sizes are as their distances from the mirror.

Supposing *a b* to be an object situated *at more than twice the focal distance* from the mirror, an *inverted and diminished* image at *A B* will correspond to it, lying between the focal point and the centre of curvature. The further the object is from the mirror the closer is the image to the focus, and the image of an indefinitely remote object, of a star for example, is situated in the focal point itself.

These images are, however, essentially different from those of plane mirrors. They are produced by the actual union *in front of* the mirror of the rays proceeding from every point of the object. They may be received upon a screen and thus be made visible on all sides by diffuse reflexion, as if the image were itself a luminous object. Such images are consequently called actual or real images. The images of plane mirrors, on the other hand, are produced by rays which *appear* to proceed from points lying behind the surface of the mirror, and are only seen when these rays pass directly into the eye. These are consequently termed *apparent* or *virtual* images.

Real images may be directly seen without any recipient screen if the observer be in the path of the rays which are again diverging after the union of the points of the image. The image appears in these cases to float in the air in front of the mirror. Aerial images of this kind produce the most surprising phenomena. For example, a beautiful bunch of flowers may be made to float over a table; it is the real image of a group of artificial flowers placed in an inverted position before a concave mirror and strongly illuminated, but concealed from the eye. If now a vase be placed upon the table in which the bunch appears to be inserted, it can easily be shown by moving the head to and fro that the bunch remains in the vase, proving therefore that the image is in front of the mirror directly above the vase.

27. Concave mirrors only furnish real images of objects which are more distant than the principal focus from the mirror. They can only give an apparent or virtual image of any object which is nearer than the

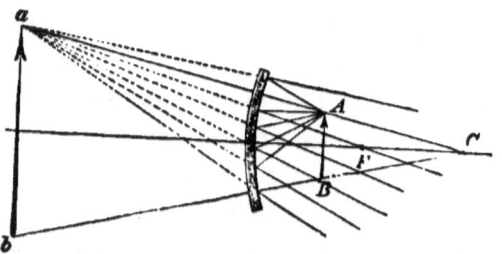

Production of a virtual image.

principal focus, because the rays of light coming from each point are reflected in a diverging manner (see fig. 28), and this image appears to an eye looking into the mirror as erect, behind the surface of the mirror, and larger than the object. Fig. 32 shows the course of the rays in the opposite case. In consequence of this enlarging action, concave mirrors are termed magnifying mirrors, and are often employed in the toilet as shaving-glasses, etc. An object placed at the principal focus of the mirror gives neither a real nor a virtual image, for the rays proceeding from each part of it are reflected parallel to their own secondary axes. If a source of light be brought into the principal focus of a concave mirror, the reflected rays proceed to great distances unimpaired in brilliancy, because they run together as parallel rays. Hence the application of concave mirrors as reflectors (Reverberen, see fig. 4, *O*) for the electric illumination of workshops during night work, and for lighthouses.

28. In spherical *convex mirrors* the reflexion takes place on the outside of the curved surface of a section of a sphere. If the aperture of the mirror be small, the rays proceeding from, or passing to, any point diverge more strongly in exact proportion as they fall on the mirror more remotely from the axis, and therefore also, after reflexion, pass through a single (real or virtual) image point.

Rays which fall parallel to a (principal or secondary) axis on a convex mirror (fig. 33) diverge after reflexion as if they proceeded from a point F, which lies on the axis about half the length of the radius of curvature behind the surface of the mirror. This may be termed the virtual principal focus. Conversely, a cone of rays converging to this point are reflected as a parallel beam. Rays which converge still more strongly, that is to say, to a point nearer to the back of the mirror, remain convergent after reflexion, and unite in a point in front of the mirror. Thus, for example, in fig. 34 the cone of rays

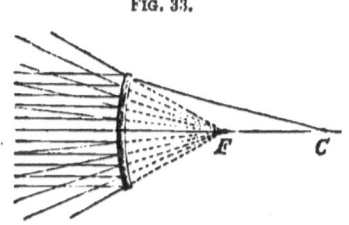

FIG. 33.

Virtual principal focus of a convex mirror.

passing to the point *b* behind the mirror, are reflected towards the point *B* in front of the mirror. If the rays proceed from a point lying in front of the mirror, they strike it divergingly, and are always reflected still more divergingly. Of any object, whatever may be its position in front of the mirror, only a *virtual erect image* can therefore be obtained, and this is perceived behind the surface of the mirror and somewhat nearer to it than the virtual principal focus (fig. 34). Since the image is always smaller than the object, a convex mirror is termed a *diminishing mirror*, and, on account of its producing pretty images, is used as a table toilet mirror.

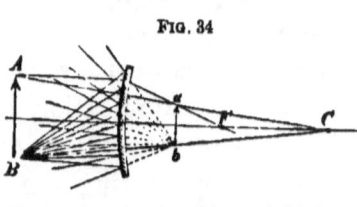

Production of a virtual image behind a convex mirror.

APPENDIX TO CHAPTER IV.

It is not difficult to deduce the propositions respecting the action of spherical mirrors of small aperture from simple geometrical considerations connected with the law of reflexion, and thus to give them a theoretical basis. Before entering upon these considerations, this opportunity may be taken of describing the best method of expressing the size of any angle. The measure of an angle is the length a (fig. 35) of the arc of a circle included between the straight lines containing the angle, drawn with a radius of any length which is taken as unity, and having its centre at the apex of the angle. Upon a second circle described with a radius $CA = \rho$, the apex

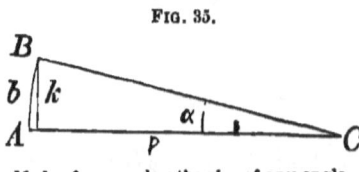

Mode of expressing the size of any angle.

SPHERICAL MIRRORS.   51

of the angle being again the centre, the same angle corresponds to the arc $AB = b$, which holds the same relation to the arc a as does the radius ρ to the radius 1. From the ratio

$$a : b = 1 : \rho,$$

however, it follows that $a = \dfrac{b}{\rho}$; that is to say, the size of any angle ACB, or the length of arc corresponding to it in a circle having a radius of 1, is always found by describing around the apex of the angle a given circle, and dividing the length of arc b between the limbs by the radius ρ.

If from the point B, where one of the limbs cuts the circle, a perpendicular k be let fall upon the second leg, this, *if the angle at C be very small*, is nearly equal to the arc b, and can be used instead of it without appreciable error. It may be admitted, that is, $a = \dfrac{k}{\rho}$ as the measure of the angle ACB. Now let ab (fig. 36) be a ray of light forming with the radius Cb

FIG. 36.

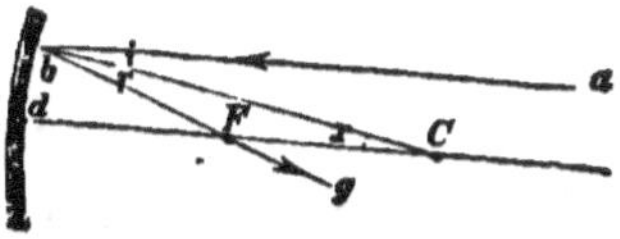

Determination of the position of the principal focal point.

(the axis of incidence), the angle i, the reflected ray bg makes with the axis of incidence the corresponding and equal angle r. The angle x, which the radius Cb and axis include, is obviously equal to the angle i, and consequently also to the angle r. Moreover, the angle bFd which the reflected ray forms with the axis is equal to the angle abF, and thus it is equal to $i + r$, or what is the same thing,

$$bFd = 2x.$$

If a perpendicular k be now conceived to fall from b upon the axis of the mirror, and if the radius of the mirror be indicated by the sign ρ, the angle x may be expressed as follows,

$$x = \frac{k}{\rho}$$

and consequently

$$bFd = 2\frac{k}{\rho};$$

and it is now clear that the angle bFd, that is to say, *the divergence of the reflected ray* from its original direction is proportional to the distance k of the point of incidence from the axis of the mirror.

The angle bFd may, however, be expressed in another way; for example, it may be said

$$bFd = \frac{k}{bF};$$

or again, because on account of the smallness of the angle bFd the line bF is scarcely different from the focal distance dF, which we indicate by f,

$$bFd = \frac{k}{f}.$$

This expression, compared with the above, leads to the equation

$$\frac{k}{f} = 2 \cdot \frac{k}{\rho},$$

which enables the position of the point F, where the reflected ray cuts the axis, to be determined. But since the magnitude k, because it appears as a factor on both sides of the equation, may be eliminated, it is obvious that the position of the point of incidence β has no influence upon the determination of the point F;

FIG. 37.

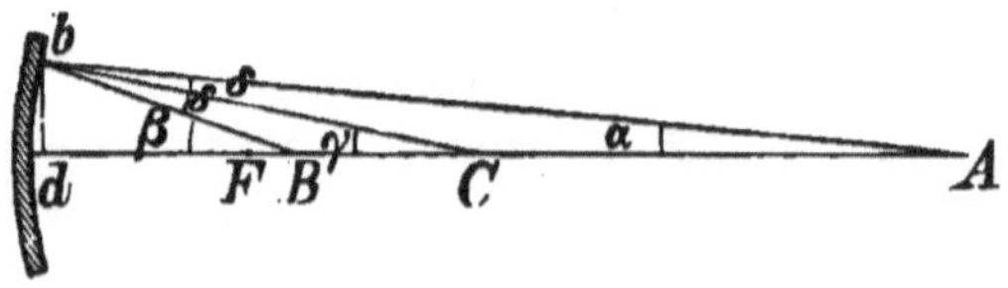

Determination of the position of conjugate points.

that is to say, *all* rays coursing parallel to the axis pass after reflexion through *one and the same point* F, situated upon the axis, the distance f of which from the mirror is determined by the equation,

$$\frac{1}{f} = \frac{2}{\rho}$$

$$f = \frac{1}{2}\rho.$$

The focal distance is consequently equal to half the radius.

If we now consider any ray $A\,b$, proceeding from the point A, making (fig. 37) the angle α with the axis, we shall find that it is so reflected in the point b that the angle of incidence and the angle of reflexion are both $= \delta$, and the reflected ray cuts the axis at the point B at an angle β. If now the angle which the axis of incidence drawn towards b makes with the axis be indicated by γ, we obtain, because β is the external angle of the triangle BCb and γ is the external angle of the triangle CAb, the two equations,

$$\beta = \gamma + \delta$$
$$\alpha = \gamma - \delta,$$

which added together make

$$a + b = 2\gamma;$$

that is to say, *for every point of the mirror the sum of the angles which the incident and the reflected ray make with the axis is inalterable, and is indeed equal to the deflection which the ray passing to the focal point experiences at the point.*

If now the focal length of the mirror be indicated by f, and its radius consequently by $2f$, and further, the distance of the luminous point dA ($= bA$) by a, the distance dB ($= bB$) of the image-point by b, and the perpendicular let fall from the point of incidence b upon the axis, by k, we obtain from the above-mentioned method of measuring the angles,

$$\alpha = \frac{k}{a}, \ \beta = \frac{k}{b}, \ \gamma = \frac{k}{2f},$$

and consequently if these are arranged in the equation $\alpha + \beta = 2\gamma$

$$\frac{k}{a} + \frac{k}{b} = \frac{k}{f}; \text{ or,}$$

since the common factor k may be eliminated,

$$\frac{1}{a} + \frac{1}{b} = \frac{1}{f}.$$

This very circumstance, that the magnitude k, which alone refers to the position, whatever that may be, of the point of incidence, is removable from the equation, supplies the proof that *all* rays proceeding from the point A, wherever they may strike the mirror, are united in the selfsame point B.

From the form of this equation, which expresses in the simplest manner the opposite relation of two conjugated points, it is further evident that the light-point and the image-point are mutually interchangeable.

The deviation which the ray incident in b experiences is 2δ. But from the above equation, it results that

$$2\delta = \beta - \alpha = k\left(\frac{1}{b} - \frac{1}{a}\right).$$

The accuracy of the statement above made, *that the deflections which the rays proceeding from any point experience are proportional to the distances of the points of incidence from the axis of the mirror*, is thus rendered evident.

In order to determine the position and size of the image by construction it is not necessary to draw a great number of rays, as in figs. 31, 32, and 34; but only two rays for each point of

FIG. '8.

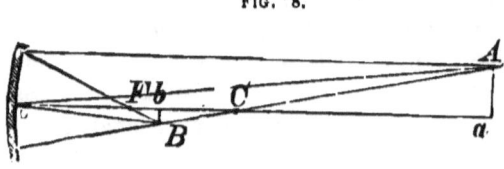

Construction showing the formation of the image.

the image, because the others necessarily meet at point where these decussate. The two rays selected should be such as to make the construction as neat and convenient as possible. In fig. 38 the object whose image is to be determined is a straight line $A\,a$, perpendicular to the principal axis. Let the secondary axis $A\,C$ be drawn to the point A; the ray coursing in this axis is of course reflected upon itself. Now let the ray parallel to the principal axis be drawn; this passes after reflexion through the principal focus, and the image of the point A required lies at the point B, where it cuts the secondary axis $A\,C$, and if $B\,b$ be let fall perpendicularly to the chief axis we obtain in Bb the image of the object Aa.

The course of all other rays proceeding from A may now be followed with facility. Thus, for example, the ray Ao, which strikes the centre of the mirror o, is reflected in the direction oB. And as at the point o the principal axis is the axis of incidence, the angle Aoa is equal to the angle Bob. If the magnitude of the object Aa be indicated by the sign p, the magnitude of the image Bb by the sign q, and the distances of the object and of the image from the mirror as before by the signs a and b, it is clear that

$$p : q = a : b;$$

that is to say, *the size of the object stands in the same relation to the size of the image as the distance of the former from the mirror is to the distance of the latter from the mirror*, a proposition that holds equally for the virtual as for the real image. The equations that have been deduced in the case of concave mirrors hold also for convex ones, if the virtual focal distance be regarded as negative, that is to say, as $-f$ instead of f.

CHAPTER V.

REFRACTION.

29. THE adjoining figure (fig. 39) represents a cubic vessel the sides of which are made of glass. A beam of parallel rays of light from the sun directed horizontally into the room by means of the Heliostat

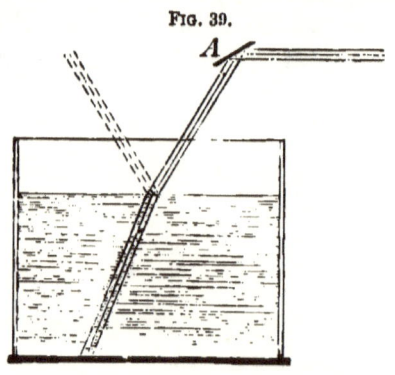

Fig. 39.

Refractor.

is thrown obliquely upon the surface of the water by a small mirror. A part of the rays is, in accordance with known laws, reflected at the surface of the water, whilst another portion penetrates it; this last, however, does not pursue a course directly continuous with the incident rays, but follows a steeper, though still always straight direction.*

---

* The course of the incident and reflected rays of light in the air is readily recognised by the illumination of floating particles of dust, and in

REFRACTION. 57

It thus appears that the rays of light, as they pass from the air into the water, are bent or refracted, and the term *refraction* is accordingly employed to indicate the phenomenon that is here observed.

The deviation of the refracted beam of light from its original direction is smaller in proportion as by turning the mirror *A* the rays are made to fall more vertically upon the surface of the water until, when they come to fall quite perpendicularly, they undergo no change of direction at all, the rays that enter the water pursuing the same direction they previously had in the air.

In order to follow the exact course of a ray of light as it passes from the air into water, or gener-

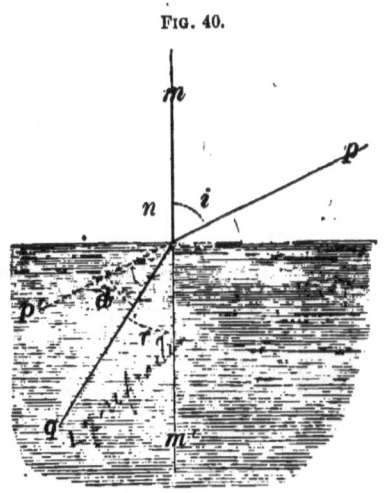

FIG. 40.

Angles of incidence and of refraction.

ally from any one transparent medium into another, let any point, *n* (fig. 40) be taken, where the incident ray

order to make it apparent in the water a small quantity of a fluorescent substance, æsculin, may be added.

strikes the surface, and upon this erect the perpendicular or axis of incidence, nm, and let this be prolonged into the second medium (nm'). We now observe, in the first place, that the *plane which contains the incident ray and the axis of incidence, always also contains the refracted ray.* It is hence termed the *plane of refraction.* The direction of the ray is determined by the angle which the ray makes with the axis of incidence, namely by the *angle of incidence i* and the *angle of refraction r.* The angle d between the refracted ray nq and the continuation np' of the incident ray, gives the amount of *deflection* which the ray undergoes in its refraction.

30. From the experiment given above it may be inferred that in the passage of a beam of light from air into water the angle of refraction is always less than the angle of incidence, and that if the angle of incidence increases, the angle of refraction and the deflection of the ray also increase. In order to obtain a more thorough insight into the whole process the relation that exists between the size of the angle of incidence and that of the angle of refraction must be investigated, and to accomplish this it is necessary to *measure* the two angles in question.

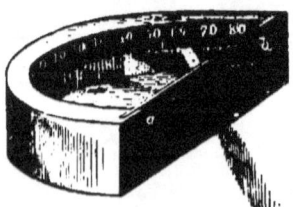

FIG. 41.

Apparatus for demonstrating the law of refraction.

Fig. 41 shows a convenient apparatus for this purpose. The flat side of a semicircular vessel is made of glass, rendered opaque except at the centre, where there is a vertical transparent slit. The internal surface of the semicircular wall is divided into 90°

REFRACTION. 59

towards each side, commencing from a point exactly opposite the slit. The vessel is half filled with water: the upper half of a horizontal beam of light, entering the vessel through the slit, pursues its original course above the level of the water, the lower half, on the other hand, experiences refraction in the water. The glass plate ab* represents the limiting refracting plane between the external air and the water, and the horizontal line drawn from the zero point of the scale to the slit, the axis of incidence. By making the vessel assume different positions in relation to the incident rays, the angle of incidence can be varied to any extent, and the angle of incidence of the ray passing over the surface of the water, and the angle of refraction of the ray passing through the water, can be read off on the scale.

We find, for example, with an angle of incidence of

15°	the angle of refraction is	$11 \cdot \frac{1}{2}°$
30°	,,	22°
45°	,,	32°
60°	,,	$40\frac{1}{2}°$
75°	,,	$46\frac{1}{2}°$

31. In accordance with this little table, the angle of incidence i being equal to 60°, the angle of refraction $r = 40\frac{1}{2}°$. If we now describe, in the plane of refraction, a circle with the point of incidence n as centre, and let fall from the points a and b, at which the incident and refracted rays cut the circle, the perpendiculars ad and bf upon the axis of incidence, it follows that bf is exactly $\frac{3}{4}$ of ad, or ad $\frac{4}{3}$ of bf. On repeating this construction for all the pairs of angles in the above

---

* It will presently be shown that this exercises no influence on the direction of the rays traversing it.

table, we constantly find *that the perpendicular corresponding to the angle of incidence is exactly $\frac{4}{3}$ as large as that belonging to the angle of refraction.* The number $\frac{4}{3}$ or $1\frac{1}{3}$, which may be regarded as the measure for the amount of refraction light undergoes in passing from air into water, is termed the *index of refraction,* or the *coefficient of refraction of water.* In passing from air into glass the rays of light are more strongly refracted, and the relation of these two perpendiculars is expressed by the fraction $\frac{3}{2}$ or 1·5. In this way every transparent substance has its own refractive power. The following table shows, in regard to a few of these, the ratio of refraction for light in passing into them from air:—

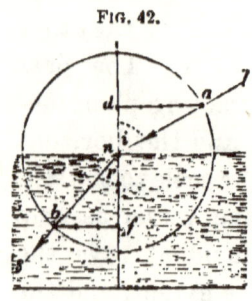

Fig. 42.

Law of refraction.

Water	1·333
Alcohol	1·365
Canada balsam . . .	1·530
Carbonic disulphide . .	1·631
Crown glass . . .	1·530
Flint glass (Fraunhofer) .	1·635
Flint glass (Merz) . .	1·732
Diamond	2·487

In geometry the perpendiculars ad and bf (fig. 42), when the radius of the circle = 1, are termed the '*sines*' of the angles i and r, and the law of refraction can be expressed in the following terms:

The sines of the angle of incidence and refraction stand in an invariable relation to each other.

If the ratio of refraction be designated by n, this

REFRACTION. 61

law can be rendered easily intelligible by the following simple expression—

$$\sin i = n \sin r,$$

that is to say, the sine of the angle of incidence is equal to n, multiplied into the sine of refraction.

If the angle of incidence be very small, by so much the smaller is the angle of refraction, for then the arcs which correspond to these angles do not materially differ from the sines, and may therefore be taken instead of them, and then the law of refraction assumes a still simpler form, namely—

$$i = n\, r,$$

that is to say, with nearly perpendicular incidence of the ray, the angle of incidence is n times as great as the corresponding angle of refraction.

32. Hitherto the passage of light from air into a fluid or solid medium where, as already stated, the re-

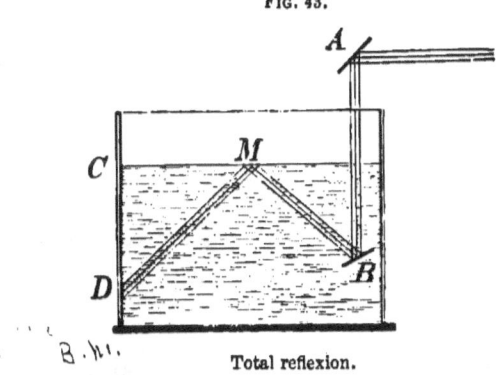

FIG. 43.

Total reflexion.

fracted ray constantly *approaches* the axis of incidence has alone been considered. In order now to acquire a knowledge of the converse, namely, of the course taken by light in passing from water into air, the cubic glass vessel (fig. 43) must again be employed, and the little mirror B which receives the beam of light directed vertically downwards by the mirror A, and reflects it upwards against the surface of the water, must be placed beneath the surface of the water. The beam, when it strikes the surface of the water at M from below, breaks up into a reflected beam which returns through the water, and into a refracted beam which passes out into the air. This last, the course of which may be easily followed both by the illuminated particles of dust in the air and by the spot of light which falls on the lid or on the opposite wall, runs in a more oblique direction than the incident beam BM. A beam of light therefore passing from water into air is thus, by refraction, deflected *from* the perpendicular; in fact, as may readily be demonstrated by measuring the angles, it follows an exactly inverse path to a ray entering water from air. Fig. 42 therefore serves to exhibit the opposite course, where bn is the ray of light which is traversing the water, and nl the ray refracted as it emerges into the air. r will of course then be the angle of incidence, and i the corresponding angle of refraction; and so it appears that if $\frac{4}{3}$ (or, speaking generally) n expresses the refraction that light undergoes in passing from air into water (or any other transparent substance) $\frac{3}{4}$ (or $\frac{1}{n}$) represents the same for the passage from water (or this other substance) into air.

By rotating the mirror B the ray BM may be made to strike more and more obliquely against the surface of the water; the emergent ray becomes similarly more and more deflected from the perpendicular, and consequently more and more approximated to the surface of the water. It is not difficult in this way to make the light spot, which enables us to follow the course of the emergent ray, strike upon the wall of the vessel towards C in the line of division between the air and the water. The emergent beam now passes along the surface of the water, and its angle of refraction amounts to 90°. It cannot, however, be refracted through an angle greater than 90°, because this is the limit of the possibility of refraction. Hence if the beam BM be directed still a little more obliquely to the surface of the water, no more light passes out into the air, the surface of the water proving absolutely impenetrable to such very obliquely falling rays. It may at the same time be remarked that at the moment when by the rotation of the mirror B the limits of refraction are overstepped and the light spot at C at the surface of the water vanishes, the ray MD, reflected inwards, which up to this time has been much fainter than the incident ray BM, *suddenly* gains in intensity and becomes just as bright as the incident ray. This is due to the circumstance that the light of the beam BM, being no longer divided into a reflected and a refracted portion, the latter is added *without loss* to the former, and the beam is said to undergo *total reflexion*. The angle of incidence at which refraction ceases and total reflexion commences is termed the *critical angle*. This amounts in the case of water to 48° 35′, for glass to 40° 49′, and for the diamond to 23° 43′. A surface at which total reflexion occurs constitutes the most perfect mirror we possess. And now let a glass prism (fig. 44) which in section forms a right-angled triangle with equal sides, be placed in the beam of light coming from the Heliostat. The rays which fall perpendicularly upon the kathetal surface AC, pass without deflection through the glass and strike at an angle of 45° (which is consequently larger than the critical angle of glass, equal to 40° 49′) upon the surface of the Hypothenuse AB. They are here totally reflected, without even a trace of light entering the air behind AB, and pass without further deflection through the second kathetal surface BC. To the eye above, the beam on its emergence is not sensibly fainter than on its entrance, and it does actually contain about 92 per cent. of the original amount of light, the loss of 8 per cent. being due to partial reflexion taking place at the surfaces of entrance and emergence. The best silvered mirrors reflect 90 per cent, mercury itself only 60 per cent, and a polished glass surface only 4 per cent. of the incident light.

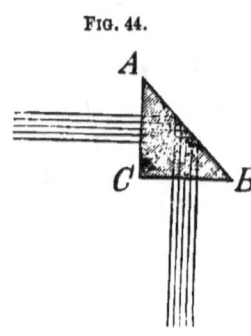

FIG. 44.

Totally reflecting prism.

33. A luminous point situated beneath the surface of the water, or more generally beneath the surface of any transparent medium, in consequence of refraction, is seen, not in the position it actually occupies, but in a *higher* position. Fig. 45 shows how the rays proceeding to the eye from the point A appear to come from the point A', which is consequently to be regarded as a virtual image of the point A. The depth of the point A' below the surface, providing the rays do not emerge very obliquely, is the nth part of the actual depth of the point A, n being regarded as the index of refraction of light in passing from air into the trans-

parent medium in question. In water, for example, all objects appear to be about one quarter less deep, hence it comes to pass that any mass of water the bottom of which can be seen, appears to be less deep

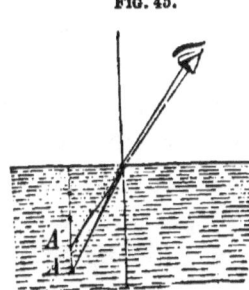

Apparent position of a point situated beneath the surface of the water.

Appearance presented by a rod dipped in water.

than it really is. For the same reason the portion of a perpendicular post which is under water appears to be shortened, and a rod held obliquely in the water to be bent at the point of immersion (fig. 46).

When the hand is dipped in water, or a coin is looked at from above, it appears to be *slightly enlarged*, because it appears to be brought nearer to the eye, and is therefore seen under a larger angle.

34. A ray of light in passing from the air, $A\ A$ (fig. 47) into a transparent medium, $B\ B$, and again emerging into air $(A\ A)$ on the other side of the medium, undergoes refraction both at the point of entrance and at that of emergence. If the ray passes through a plate bounded by parallel surfaces, it becomes, as is shown in fig. 47, approximated to the axis of incidence at the point of

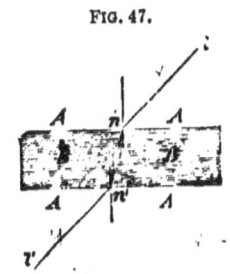

Refraction through a transparent plate with parallel surfaces.

entrance, and diverted from it to the same extent at the point of emergence. The emergent ray consequently pursues its course *parallel* to the entering ray, though without forming its direct continuation. The only change it undergoes from its original direction is a lateral shifting, which is greater in amount the more obliquely the ray strikes the plate, the thicker the plate, and the greater its index of refraction. Thin plates, as for example the ordinary panes of glass in our windows, produce so slight a shifting that objects are seen through them of their ordinary size and shape, and in their natural position. That a ray of light, after its passage through a plate with parallel surfaces continues to pass in a direction parallel to its original direction, and only undergoes a lateral shifting, may be easily demonstrated by a simple experiment. If a thick plate of ordinary glass be held in a beam of light proceeding from the mirror of the Heliostat so that about half the beam passes without obstruction at the side of the plate whilst the other half is refracted through it, it will be seen that the latter portion continues parallel to the former and throws a light upon a screen placed opposite to it, which is more distant from the light thrown by the direct rays in proportion as the rays are made to strike the plate more obliquely. Let a second plate of flint glass be now placed upon the first plate; the lateral shifting increases, but the emergent rays still always remain parallel to the entering rays, nor is any change in the parallelism produced if a third plate be added. However numerous may be the transparent plates of different substances superimposed on each other, the rays on their emergence into the air remain parallel to their course in the air before their entrance into the transparent medium.

REFRACTION. 67

Now since in the passage of a ray of light through the two plates A and B (fig. 48) the angle of emergence i' is equal to the angle of incidence i, the refracted ray must pursue the same course in the medium B which it would have had if this medium had been struck directly by the incident ray passing in the direction i, after

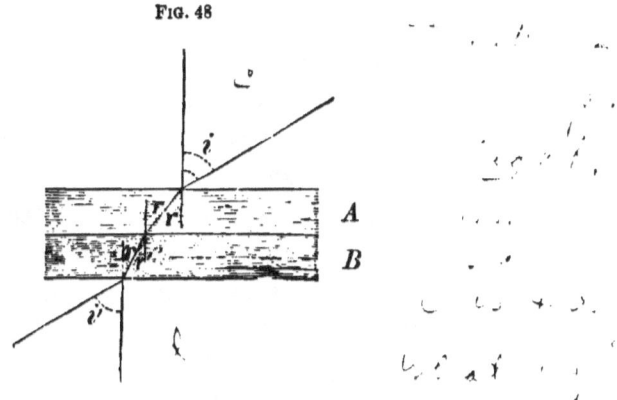

FIG. 48

Refraction through two parallel plates.

removal of the plate A. *The plate A therefore exerts no influence upon the direction of the rays of light in the medium B.* It is now obvious that in the experiment described in § 30, the glass plate ($a\,b$, fig. 41) through which the rays must pass before they penetrate into the interior of the vessel, does not interfere with the result because it does not cause any alteration in the direction of the refracted ray.

From the circumstance that a pencil of light in traversing two or more parallel plates undergoes no change in direction, it is moreover legitimate to conclude[*] that the index of refraction of a pencil of light in passing from one medium, A, into a second

---

[*] See Appendix to this Chapter.

medium, B, may be expressed by the quotient $\frac{n''}{n'}$, where n'' represents the index of refraction of the medium B, and n' that of A in relation to the air. Thus, for example, the index of refraction from water into glass $= \frac{1\cdot530}{1\cdot333} = 1\cdot148$.

35. When a beam of light traverses a transparent body, the opposite surfaces of which are inclined to one another, the emerging ray no longer remains parallel to the incident, but is diverted from its original direction, and fig. 49 shows the course of the beam under these circumstances. A straight triangular prism of glass (fig. 50) may be used for experiments on this mode of deflection. When the surfaces $a\,b\,c\,d$ and $a\,b\,g\,f$ are used as surfaces of entrance and emergence, the edge, $a\,b$, in which these two surfaces meet is termed the refracting edge, and the angle, $d\,a\,f$, where they meet, the *refracting angle* of the prism. All planes which, like the terminal surfaces $d\,a\,f$ and $c\,b\,g$, or planes parallel to them, are perpendicular to the refracting edge, are termed *chief* or *principal sections* or *planes* of the prism, and the remarks here made will be limited to those rays which run in principal sections.

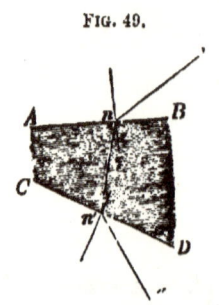

FIG. 49.

Refraction through a piece of glass, the surfaces of which are not parallel.

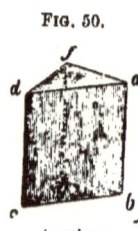

FIG. 50.

A prism.

If the opening of the Heliostat be closed with a red glass,* and a prism (fig. 51) with vertically-placed refracting edge be brought in the path of the horizontal red pencil of

---

* The object of this proceeding will be presently explained.

REFRACTION. 69

light, so that about one half of the rays passing by the side of the edge, *A*, pursue their original direction, *A D*, whilst the other half are refracted by the prism and deflected towards *A E*; the amount of deflection, that is to say, the size of the angle *D A E* between the emergent and the direct rays, will be found to vary as the position of the prism in regard to the incident rays,

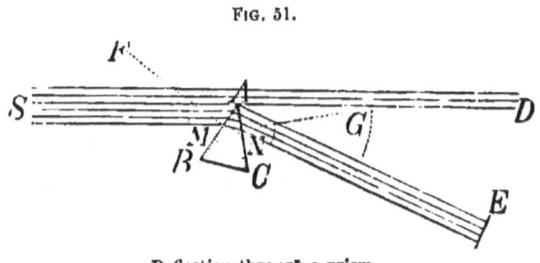

Fig. 51.

Deflection through a prism.

or, which comes to the same thing, as the direction of the rays in relation to the prism is altered.

On rotating the prism to a greater or less extent, a position may easily be discovered in which the deflection is less than in any other position. As it is turned away from this position in either direction, or, which expresses the same thing in other words, as the rays are made to fall more or less obliquely upon the prism than in the position of least deflection, the deflection becomes constantly more and more marked.

In order to determine the course pursued by a ray of light with the least deflection, the following experiment may be made. A part of the incident light is reflected at the anterior surface, *A B*, of the prism, towards *M F*. The half of the angle, *S M F*, is consequently the incident angle. If a small mirror be placed at *E*, perpendicularly to the emergent rays, these are reflected back upon themselves, and are reflected at the posterior surface, *A C*, of the prism, towards *N G*; then the half of the angle *E N G* is the emergent angle. It may now be easily shown by direct measurement that if the prism be placed in the position of least deflection, the angle *S M F* is equal to the angle *E N G*, or that the angle of entrance and of emergence are equal to each other. But if the incident and the emergent rays form equal angles with the surfaces of the prism, the refracted ray *M N*, in its course through the prism, must be equally inclined to both surfaces. *The minimum deflection occurs therefore when the ray in the interior of the prism forms equal angles with the surfaces of entrance and of emergence.* The knowledge of the minimum deflection of a prism is a matter of great importance in practical optics, because we are able from it and the refracting angle of the prism to determine with great exactness the index of refraction of the substance of which it is composed.

From fig. 52, which represents the course of a ray

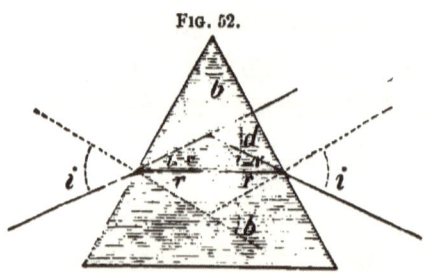

Smallest deflection through a prism.

of light in the case of least deflection, it results* that the angle of refraction, *r*, is equal to half the angle of

---

* See Appendix to this Chapter.

the prism b, and the angle of incidence, i, is equal to half the combined minimum deflection and prism angle.

If, however, the angle of refraction belonging to the angle of incidence, i, be known, the index of refraction must, in accordance with the law of refraction, be equal to the ratio between the sines of these two angles.

In order to obtain the index of refraction of a body, the following method is adopted. A prism of the substance is prepared, the refracting angle of which is measured by the reflecting Goniometer (§ 19), and the minimum deflection is determined when, by testing, it has been brought into the right position. From these two data, which can be ascertained with great accuracy, the index of refraction can be easily deduced by the above method.

In order to give to a fluid the form of a prism it is introduced into a vessel in which the opposite inclined walls are made of plates of glass, carefully ground to plane surfaces. Fig. 53 is such a hollow prism. As plates with parallel surfaces do not alter the direction of the rays of light, they do not interfere with the measurement of the deflection caused by the fluid.

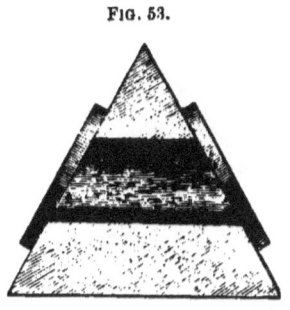

Fig. 53.

Hollow prism.

The indices of refraction given above (§ 31) were all obtained in this manner.

36. When a comparison is made of several prisms composed of the same kind of glass, the refracting angles of which differ, it is found that the minimum deflection increases more quickly than the refracting angle. Thus for prisms of ordinary glass it appears that,

When the refracting angle amounts to 20°, the minimum deflection amounts to 10° 49'.

When the refracting angle amounts to 40°, the minimum deflection amounts to 23° 6'.

When the refracting angle amounts to 60°, the minimum deflection amounts to 39° 49'.

It is only in the case of prisms with very small refracting angles that the deflection holds the same ratio, for it is found that

With a refracting angle of 2° the minimum deflection is 1° 3½'.

With a refracting angle of 4° the minimum deflection is 2° 7⅓'.

With a refracting angle of 6° the minimum deflection is 3° 11'.

The amount of refraction in thin acute-angled prisms does not alter materially even if the incident ray is inclined several degrees to that which traverses the prism under equal angles. For example, the prism of 4° may be moved as much as 5° to one side or the other from the position of minimum refraction, or may thus be rotated 10° without the deflection varying more than a minute.

It may therefore be laid down that *a prism with very small refracting angle,* as long as the rays do not fall too obliquely upon it, *invariably produces an amount of deflection proportional to the refracting angle.*

APPENDIX TO CHAPTER V.

To §§ 31 and 32. By means of the law of refraction the angle of refraction corresponding to each angle of incidence (and the converse) may be easily determined either by calculation or by construction. The latter may be conducted in the mode indicated in fig. 42. The construction shown in fig. 54 is still more convenient. Two circles are described around the point of incidence in the plane of refraction, one of them with a radius = 1, the other with the radius=n, n being the index of refraction of the ray in passing out of the first into the second medium. Now let the incident ray $l\,n$ be prolonged to intersect the first circle in the

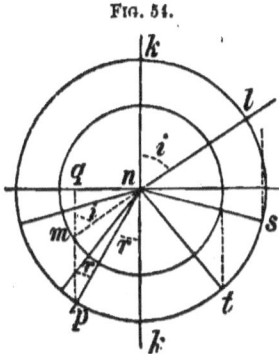

Fig. 54.

Construction of the refracted ray.

point m, and through m drawn $p\,m\,q$ parallel to the axis of incidence, intersecting the second circle in the point p, then $n\,p$ is the direction of the refracted ray. For since the angle $q\,m\,n$ is equal to the angle of incidence i, the sin $i = q\,n$; and further, since the angle $q\,p\,n$ is equal to the angle r, n sin $r = q\,n$, and hence as is required by the law of refraction,

$$\sin i = n \sin r.$$

For any ray $p\,n$ proceeding from the second medium, let a line parallel to the axis of incidence be drawn through p to cut the first circle at the point m, then the line $m\,n$ produced, gives the direction of the emerging ray $n\,l$.

The last construction becomes *impossible* when as in the ray $s\,n$ the parallel to the axis of incidence no longer cuts the first circle. The total reflexion which this ray experiences is thus rendered intelligible.

If the parallel touches the first circle just at the end of its horizontal diameter, as occurs with the ray $t\,n$, the refracted ray passes out towards $n\,q$ along the limiting surfaces of the two media, and $t\,n\,k' = \gamma$ is the critical angle. The ratio thus holds, as appears from the construction

$$n \sin \gamma = 1, \text{ or } \sin \gamma = \frac{1}{n}.$$

To § 34. That the index of refraction in passing from a medium A into a second medium B, is equal to the quotient $\dfrac{n''}{n'}$, where n' represents the index of refraction of the medium A, n'' that of the medium B, as compared with air, can be demonstrated in the following manner. In fig. 55, which represents

Fig. 55.

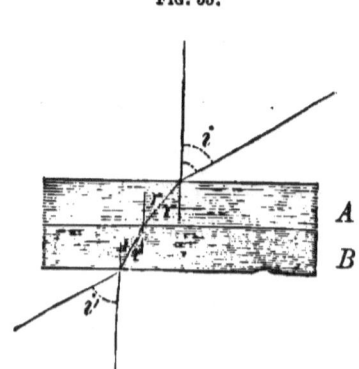

Refraction through two parallel plates.

the passage of a ray of light through two parallel plates, we have on entrance into the first plate

$$\sin i = n' \sin r,$$

and on emergence of the ray from the second plate into the air

$$\sin i' = n'' \sin r'.$$

But inasmuch as the emergent ray is parallel to the incident ray, $i = i'$ consequently also, $\sin i = \sin i'$, and

$$n' \sin r = n'' \sin r'$$

or

$$\sin r = \frac{n''}{n'} \sin r'.$$

In the transition of the ray from the first into the second plate, r is obviously the angle of incidence, and r' the angle of refraction, and consequently $\frac{n''}{n'}$ is the ratio of refraction corresponding to this transition.

To § 35. The deflection of the incident ray caused by a prism placed in any given position amounts to the sum of the deflection on entrance and the deflection on emergence. If i and i' (fig. 56) indicate the angles which the incident and the emergent ray, and r and r' the angles which the ray in its course through the prism makes with the axis of incidence, then $i-r$ is the amount of deflection in the first, and $i'-r'$ that in the second refraction. The total deflection, D, as appears from the figure, is the sum of the two separate deflections, so that

$$D = i - r + i', - r' \text{ or } D = i + i' - (r + r').$$

From the figure it may also be concluded that the sum of the two angles o refraction remains constantly equal to the refracting angle of the prism b, or that constantly

$$r + r' = b.$$

Consequently the deflection may also be expressed in the following form:—

$$D = i + i' - b.$$

When in the case of the minimum refraction (d, fig. 52), $i = i'$, and $r = r'$, we obtain

$$2r = b \text{ and } d = 2i - b.$$

Thence it results that the angle of incidence $i = \frac{1}{2}(d + b)$, and the angle of refraction $r = \frac{1}{2} b$. We obtain therefore for the calculation of the index of refraction the equation

$$n = \frac{\sin \frac{1}{2}(d + b)}{\sin \frac{1}{2} b}.$$

That the minimum of deflection occurs with equiangular transit, *i.e.* when the ray of light makes equal angles with the two sides of the prism, may be shown by the following statement:—We consider that the course of any ray of light in the prism is as in fig. 56, from left to right and upwards; we compare

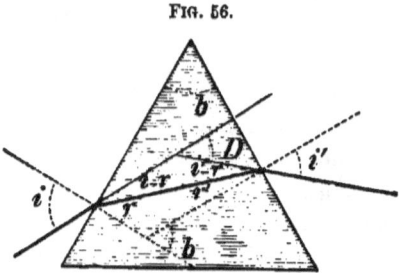

Passage of a ray of light through a prism.

with this a second ray, which runs with equal inclination to the two surfaces from the left to right, and downwards; these two rays lie symmetrically with regard to the equiangular ray of fig. 52, and undergo, since they only in this respect differ from one another, that i and i', and also r and r', are interchanged, equal amounts of deflection. It may now be easily shown that the amount of deflection of the non-equiangular ray of fig. 56 is greater than that of the equiangular ray of fig. 52.

The angle r in fig. 56 is greater than with equiangular rays, the angle r' on the other hand is just as much smaller, since the sum of $r + r' = b$. If we proceed consequently from equiangular to non-equiangular rays the angle i augments, whilst i' diminishes. By means of the construction fig. 54 it may easily be demonstrated that if the angle of refraction r be allowed to increase and diminish about equally, the increase of the angle of incidence i is in the former case *greater* than is its diminution in the latter. In the transition from equiangular to any other rays consequently, in the expression

$$D = i + i' - b,$$

the angle i augments so much the more as the others diminish: that is to say, the deflection of the ray becomes *greater*, or which is the same thing, the minimum deflection occurs with equiangular transit.

To § 36. The proposition laid down in § 36 in regard to acute-angled prisms may be easily established theoretically. If for example the refracting angle of a prism be very small, those rays which are near to the minimum deflection deviate but little from the axis of incidence. Here, therefore, only very small angles of incidence and emergence are dealt with, to which the simplified law of refraction applies (see end of § 31), from which it appears that

$$i = nr \text{ and } i' = nr'$$

and the deflection

$$D = n(r + r') - (r + r') = (n - 1)(r + r')$$

or because

$$r + r' = b,$$
$$D = (n - 1) b,$$

that is to say, the deflection, whatever may be the angle of incidence, providing only that it remains very small, is determined exclusively by the index of refraction and the refracting angle of the prism, and indeed is proportional to this last.

CHAPTER VI.

LENSES.

37. ~~Thin~~ pieces of glass, (sometimes thin) the two surfaces of which (or one surface, the other remaining flat) have been ground to a spherical form, are termed lenses.

Convex lenses are those which are thicker in the middle than at the edge. Fig. 57 exhibits three different forms, as seen in section, namely, *a* a *bi-convex*, *b* a *plano-convex*, and *c* a *concavo-convex* lens.

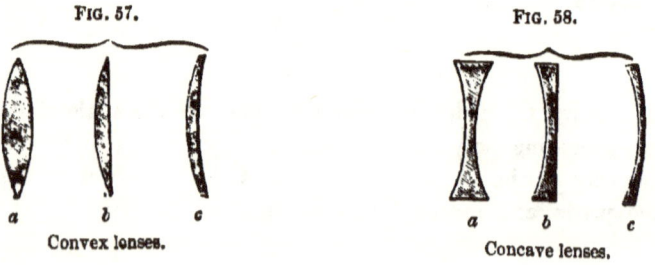

Fig. 57.   Fig. 58.

Convex lenses.   Concave lenses.

Concave lenses (fig. 58) are thicker at the edges than in the middle: *a* is a *bi-concave*, *b* a *plano-concave*, and *c* a *convexo-concave* lens.

The term axis of a lens indicates the straight line which joins the centres C and C' (fig. 59), of the two spheres of which the limiting surfaces are segments. Where one of the surfaces is flat, a line drawn perpendicularly to that surface from the centre of curvature of

LENSES. 79

the curved surface is regarded as the axis. The form of a lens is symmetrical around its axis, for all planes passing through the axis, which are termed chief or principal planes or sections, have the same sectional outline.

The angle $A\,C\,B$ (fig. 59) which two straight lines, drawn to diametrically opposite points of the border of

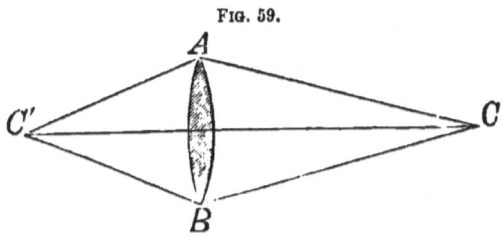

Axis and centres of curvature.

the lens from the centre of curvature, make with one another, is termed the *aperture of the corresponding surface of the lens*. We shall here only have to do with such lenses as have a small aperture not exceeding six or eight degrees at most.

38. If a pencil of parallel rays from the sun be directed upon a bi-convex lens (fig. 60), parallel with its axis, these will be so refracted that they will all pass through one and the same point, F, situated on the axis on the other side of the lens, which is called the focus.

If the several rays be followed in their passage through the lens it is observable that each is refracted in exactly the same mode as in a prism whose refracting angle is turned away from the lenticular axis, with this difference, however, that for each ray there is a different refracting angle. The small angle between the directions of the two lenticular surfaces at the points of entrance and emergence of the ray in question is to be regarded as the refracting angle. This angle is proportionally greater as we recede from the axis of the lens.* The lens acts just as if each ray struck an acute-angled prism, the refracting angle of which is

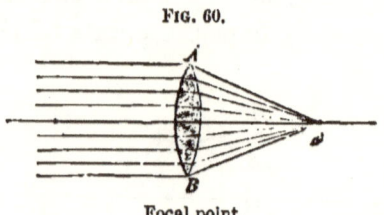

Fig. 60.

Focal point.

greater in proportion as the point of incidence is further from the axis.

If what has been said above in regard to the relation of acute-angled prisms be remembered, it may be conceived that rays pursuing a parallel course on this side of the lens, the further they severally strike the lens from its axis, must run together on the other side of the lens into one and the same point of the axis. The ray which runs in the axis itself meets parallel surfaces upon its entrance and emergence from the lens, and therefore experiences no deflection.

On the supposition that the rays fall parallel upon the surface of the lens from the side towards F, their union will then occur on the other side of the lens in a point of the axis which will also be at the same distance from the lens as the point F, because the rays will meet the same refracting angles at the same distance from the axis, and will consequently experience the same deflection as before. Every lens therefore possesses *two* focal points upon its axis, which are placed on opposite sides of it, at the same focal distance.

* See Appendix to this Chapter.

LENSES. 81

39. The flame of an electric lamp is now to be brought into the focus F (fig. 60) of the lens. The result may be predicted. A beam of light, composed of rays running parallel to the axis, emerges on the other side of the lens. Following the plan previously adopted, it may be said that rays proceeding from the focus on one side of the lens run on the other side towards an infinitely remote point of the axis.

If the light from the lens be now removed till it reaches the point R (fig. 61), a cone of rays may be seen to emerge which converge towards a point S on the axis. This point S, in which all the rays proceeding from R that fall upon the lens unite, is the *real image* of the luminous point R.

When the luminous point R (fig. 62) is brought to exactly double the focal distance from the lens, its image, S, on the other side, will be double the focal distance from the lens also.

When the luminous point is placed at S (fig. 61) its image is formed at the point R, which was before the

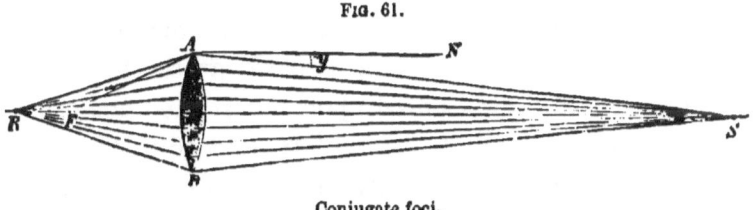

Fig. 61.

Conjugate foci.

position of the light. The points R and S are consequently so associated, that the one is the image of the other, and they are said to be *conjugate* to each other. When one is more than double the focal distance from the lens, the other is less upon the opposite side, but always at *a greater distance* from it than the simple focal distance.

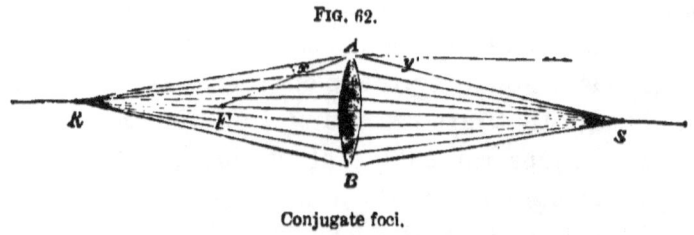

Fig. 62.

Conjugate foci.

If the luminous point T (fig. 63) be situated between the focus and the lens, this no longer has the power of

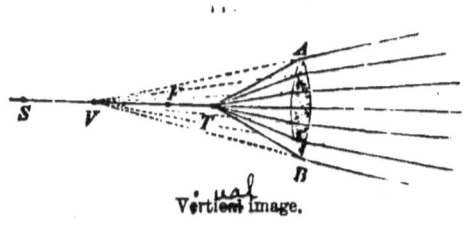

Virtual Image.

making the strongly divergent rays parallel or convergent, but simply diminishes their divergence. An actual union of the refracted rays can now no longer take place, but if prolonged backwards, they pass through a point V, situated upon the axis on the other side of the lens, which is more remote from the lens than the luminous point T; in other words, the rays emanating from T proceed divergingly after having traversed the lens, just as if they emanated from the point V. The point V is consequently the virtual image of the point T.

If, conversely, a converging pencil of rays proceeding from the right side (fig. 63), fall upon the lens which is directed to the virtual luminous point V, the rays are made to unite at the real image-point *T*. The points *T* and *V* therefore constitute image-points of each other, and they also are consequently termed *conjugate* points.

40. The behaviour of lenses, in regard to light, which has just been described, is easily explained by the peculiarity that prisms with small refracting angle possess of deflecting equally all rays, whatever may be their direction, providing they do not fall too obliquely upon them. In consequence of this peculiarity, all rays which are not inclined to the axis at too great an angle must undergo the same deflection at one and the same point of the lens. The ray *R A*, for example (fig. 61), striking near the edge of the lens, inasmuch as it is refracted towards *A S*, undergoes the same deflection which the ray *A N*, running parallel to the axis, experiences; that is to say, the angle *R A S*, wherever the luminous point *R* may be, is always equal to the angle *F A N*, the magnitude of which is given, once for all, with the focal distance. The conjugate points may be very easily determined by construction; if the angle *F A N* be cut out of a piece of cardboard, and having been placed with its apex upon the point *A* and rotated around this point, the sides containing the angle *T* then always cut the axis in two points conjugate to each other.

It results as a necessary consequence from the above-mentioned proposition, according to which in lenses of small aperture the deflection of a ray is greater in proportion as the part of the lens which it strikes is further from the axis, that all rays proceeding from any point on the axis pass again after refraction through some point of the axis.

84   OPTICS.

41. The concordance which exists between the properties of convex lenses, so far as we have at present gone, and those of concave mirrors, is so remarkable that it is scarcely necessary that they should be expressly pointed out. And it will not be surprising if in the course of the following researches results are obtained essentially agreeing with those already given in the case of concave mirrors.

If, for example, I place the light of an electric lamp at a (fig. 64) above the axis, its image is formed

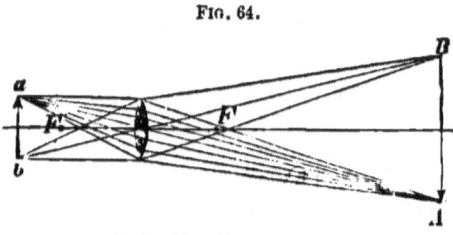

Fig. 64.

Production of a real image.

below the axis in A. An imaginary straight line joining the points a and A passes through the centre O of the lens, and a ray striking the lens in this direction ($a\,O$) undergoes no deflection, because it meets parallel portions of the surface of the lens. It behaves itself consequently like a ray running in the axis itself. The term *secondary* axis has therefore been applied to every line passing through the centre of the lens, in order to distinguish such lines from the chief axis which joins the centres of the two spheres of which the surfaces of curvature are segments. The same laws hold in regard to each secondary axis for rays that do not fall too obliquely, as has already been stated as applying to the chief axis. A pencil of rays, for example, which falls upon the lens parallel to its secondary axis $a\,O$, will be united in a point upon this secondary axis at about the focal distance of the lens $O F$. Every secondary axis consequently also possesses two focal points, and its conjugate points are in all respects similar to those of the chief axis.

42. If from the points a and A, which correspond as light object-point and image-point on the secondary axis $a\ O\ A$, we let fall the lines $a\ b$ and $A\ B$ perpendicular to the principal axis, so that each is bisected by the chief axis, the points b and B upon the secondary axis, $b\ o\ B$, are obviously also conjugate to each other. So long as the angle between the secondary axis $a\ O$ and the principal axis is very small, all points of the line $a\ b$ may be regarded as equally remote from the middle of the lens O, and likewise all points of the line $A\ B$. Every point of the line $a\ b$ has therefore its conjugate point upon the line $A\ B$, which is at the spot where these are struck by their own axis. The middle points of $a\ b$ and $A\ B$, for example, are conjugate points upon the chief axis.

From the preceding illustration, which is limited to the plane of the construction, a more general statement affecting the space around the chief axis can easily be deduced. *If, for instance, the vertical planes $a\ b$ and $A\ B$ be conceived to be placed at two conjugate points of the chief axis, each point of the one plane will have its image in the other plane at the spot where this is struck by the axis corresponding to each point.* The two planes are said to be ' conjugate to each other.' If therefore any line be situated in the one plane, there is projected from the lens an exact image of it upon the other conjugate plane, the size of which is in the same proportion as their relative distances from the lens. And what has here been stated in regard to a flat figure holds also for any material object the parts of which do not project too far beyond a plane perpendicular to the chief axis.

As long as the object is situated at a greater distance from the lens than the focal distance, an actual reunion of the rays of light occurs upon the other side in the image plane; and thus an *actual* or *real* image is formed, which may be received upon a screen and thus made objectively apparent. The real images are of course always inverted in relation to the object.

It is easy to show this relation by experiment. Let a lighted candle be placed in front of a lens (fig. 65), and somewhat beyond its focal distance, by a little

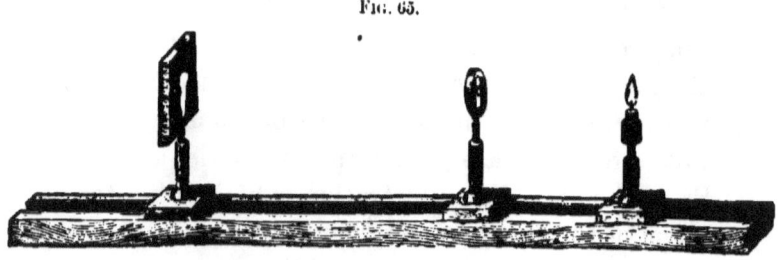

Fig. 65.

Real image seen through a convex lens.

shifting to and fro of the screen, the place of the image may be easily determined, and it will be found that it is situated a little beyond twice the focal distance, and that it is inverted and enlarged. If the position of the screen and candle be so altered that the candle is situated a little beyond, and the screen a little nearer than twice the focal distance of the lens, an *inverted diminished* image of the flame is obtained upon the screen. Fig. 64 exhibits the course of the rays in both cases; if $a\,b$ be the object, $A\,B$ is its real image, and *vice versâ*.

43. If an object be situated at somewhat less than

the focal distance from the lens, no real image of it can be projected by the lens. For the rays which emanate from one of its points (*A*, fig. 66) will now no longer be collected into one point on the other side, but issue

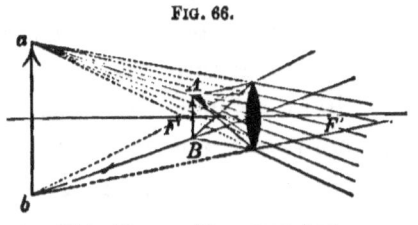

Fig. 66.

Virtual image with a convex lens.

divergingly from the lens, just as if they came from a point *a* situated on the same side of the lens but more distant from it than the point *A*. An observer looking through the lens from the other side sees therefore instead of the small object *A B*, the *enlarged virtual image a b*, which is erect in regard to the object. On account of this well-known action, convex lenses are called *magnifiers*. Every lens specially destined for this object of enabling us to see the enlarged virtual images of small objects is called a magnifying glass (Lupe).

44. A concave lens acts at each part like an acute-angled prism, the refracting angle of which is turned towards the principal axis, and is greater the further the point is from the axis. Every ray that strikes such a lens will therefore be turned away from the axis, and to a greater extent in proportion as the part of the lens on which it falls is further from the axis. Hence the solar rays, which are directed upon this biconcave lens (fig. 67), parallel to its axis, issue divergingly from the other side of the lens in such a manner that

they appear to proceed from a point F, situated upon the axis on this side, which we may designate as the *apparent* or *virtual focus*. Every concave lens has, on every axis, two such focal points, which are situated at an equal distance from the lens on either side, and have the same significance as the real foci of a convex lens. The virtual focal distance is proportional to the deflection from the axis which the rays of light experience at each point of the concave lens.

In order to elucidate the action of concave lenses, a precisely similar series of observations to those already given in the case of convex lenses should here be inserted; but to avoid repetition it will be sufficient if the more important cases are here mentioned.

A cone of rays, produced by a convex lens, is allowed to fall upon a concave lens (fig. 67) in such a manner that the rays converge towards its focal point F on the opposite side; in this case there proceeds from the other side of the lens a cylinder of rays parallel to its chief axis. If the incident rays converge to a point which is more distant on that side than the focus of the lens, they must emerge divergingly: but if they converge to a point B, situated nearer to the lens (fig. 68), they must converge after refraction less strongly to the more remote point A. Rays, lastly, which are emitted divergingly from a point A, as, for example, from an electric lamp placed at this spot, are rendered still more divergent by the lens, as if they proceeded from a point B situated nearer to the lens on the same side.

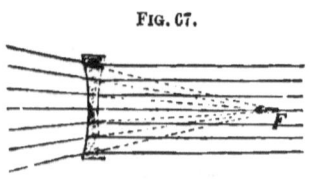

FIG. 67.

Virtual focus of a concave lens.

Hence it follows that a concave lens can form a virtual image only of an object, whatever may be the distance that this is from it, because it makes the di-

FIG. 68.

Action of a concave lens on convergent and divergent rays.

verging rays emitted from every point of the object still more divergent. The eye of an observer looking through the lens (fig. 69) receives the rays emitted from the object AB as if they came from a *diminished erect virtual* image ab. On account of this diminishing action, concave glasses are called *diminishing* glasses, and thus we see, that whilst convex lenses are analogous

FIG. 69.

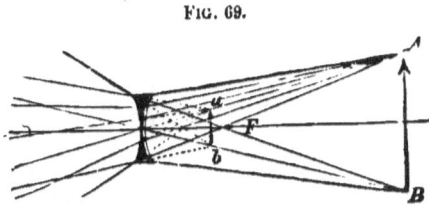

Virtual image formed by a concave lens.

to concave mirrors in their action, concave lenses correspond to convex mirrors.

45. Of the various forms of lenses enumerated in § 37 we need only consider the biconvex and biconcave more closely, because the remaining forms entirely agree in their action with these representatives of these groups.

The lenses of the first group possess *real* foci; they make parallel incident rays convergent, and unite them into one point; they make convergent rays still more convergent, divergent rays less divergent, or even convergent.

The lenses of the second group have *virtual* foci; they make parallel rays divergent, divergent still more divergent, convergent less convergent, or even divergent.

Every lens which becomes thicker towards its periphery has virtual foci; and *vice versâ*, for the focus of a lens to be real the lens must be thicker in the middle than at the edge.

For all lenses, however, to whatever group they may belong, the general statement holds good that rays, which before they strike upon the lens pass through a single point, pass also, after refraction, through a single point which is conjugate to the first, upon the axis passing through it.

APPENDIX TO CHAPTER VI.

To § 38. The angle which the anterior surface of a lens fig. 70) makes at the point K, which is distant $KP = k$ from

FIG. 70.

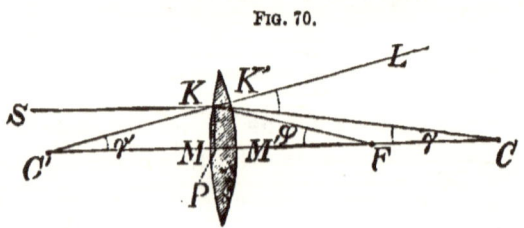

Determination of the focal distance.

the axis, with the posterior surface of the lens at the opposite point K', or in other words the refracting angle corresponding to the point K, is equal to the angle CKL, which the radii CK and $C'K'$

LENSES.

prolonged to K from the centres of curvature C and C' form with each other, because these radii are obviously perpendicular at K and K' to the surfaces. The angle CKL, however, as external angle of the triangle CKC', is equal to $\gamma + \gamma'$. Presupposing that the lens is one of small aperture, the angles above named, as well as those more distant, are collectively very small, and in order to express them we may use the method applied above. (See Appendix to Chapter IV.)

Consequently

$$\gamma = \frac{k}{CK} \text{ and } \gamma' = \frac{k}{C'K'}.$$

If, as always occurs in ordinary cases, the thickness of the lens is very inconsiderable as compared with its radius of curvature, we may, without risk of material error, take $C'K'$ instead of $C'K$. If therefore we indicate the radii CK and $C'K'$ respectively, by r and r' we obtain

$$\gamma = \frac{k}{r} \text{ and } \gamma' = \frac{k}{r'}.$$

The refracting angle $\gamma + \gamma'$ at the point K is therefore

$$k \left(\frac{1}{r} + \frac{1}{r'} \right),$$

that is to say, it is proportional to distance k from the axis.

We now know (see Appendix to Chapter V.) that the deflection produced by an acute-angled prism is equal to $(n-1)$ times its refracting angle. Every ray falling upon the lens at the point K undergoes therefore the deflection

$$k(n-1)\left(\frac{1}{r} + \frac{1}{r'}\right).$$

The ray SK, for instance, which runs parallel to the axis, since it is deflected to the focal point F, undergoes a deflection that is represented by the angle φ, which the refracted ray forms with the axis. From what has just been stated,

$$\varphi = k(n-1)\left(\frac{1}{r} + \frac{1}{r'}\right).$$

The angle ϕ may also be expressed by

$$\phi = \frac{k}{FK}.$$

FK, however, if the small thickness of the lens be neglected, may be replaced by FM, that is to say, by the focal distance f of the lens, so that we get

$$\phi = \frac{k}{f}.$$

If this value be substituted for ϕ in the above equation, the factor k, common to both sides, may be eliminated, and we obtain for the calculation of the focal distance the equation,

$$\text{I.} \quad \frac{1}{f} = (n-1)\left(\frac{1}{r} + \frac{1}{r'}\right).$$

The very fact that k is eliminated from the equation demonstrates that all rays parallel to the axis, at whatever distance k from the axis they may fall upon the lens, unite on the other side in the single point F.

It appears, further, from the circumstance that the radii r and r' can be substituted for each other without altering the expression for the focal distance, that the focal distance is equal for the two sides of the lens.

The formula shows also in what way the focal distance is dependent upon the index of refraction n of the substance of which the lens is composed. For a biconvex lens composed of crown glass ($n = 1.530$), for example, the two radii of curvature of which are equal, $r' = r$, we find

$$\frac{1}{f} = 0.53 \cdot \frac{2}{r} = \frac{1.06}{r},$$

consequently

$$f = \frac{r}{1.06} \text{ or } f = 0.943 \cdot r$$

With a biconvex lens of crown glass of equal curvature on both sides, the focal distance is consequently nearly equal to the radius of curvature, that is to say, the focus is very nearly coincident with the centre of curvature. For a similar lens composed of flint glass ($n = 1\cdot635$), it results on the other hand that

$$f = 0\cdot787 \cdot r,$$

and for a lens composed of Diamond

$$(n = 2\cdot487)$$
$$f \text{ only} = 0\cdot336 \cdot r.$$

From this it is evident that for lenses of similar form, but made of different materials, the focal distance becomes smaller as the index of refraction of the substance used increases.

To § 39. In order to determine the position of the conjugated points, it is only necessary to follow any given ray in its course. For this purpose we select a ray, RA (fig. 71), striking the border of the lens, which is refracted in the line AS, so that

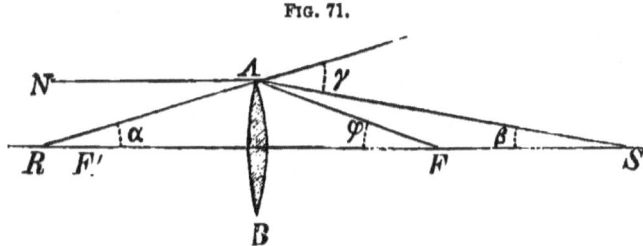

FIG. 71.

Determination of conjugate points.

R and S are conjugate points. The deflection, γ, which this ray undergoes at A is the same in amount as the deflection ϕ which the ray NA parallel to the axis experiences at the same point; that is to say, $\gamma = \phi$. But if the angles which the rays RA and AS make with the axis be indicated by a and β, $\gamma = a + \beta$. It results consequently that

$$a + \beta = \phi.$$

If the distance of the point R from the lens be indicated by a, that of the point S by b, the focal distance by f, and lastly, the distance of the point A from the axis by k, the equations

$$a = \frac{k}{a}, \ \beta = \frac{k}{b}, \ \phi = \frac{k}{f},$$

are obtained, and since also

$$\frac{k}{a} + \frac{k}{b} = \frac{k}{f},$$

from which equation the magnitude of k which refers to the several points of incidence may be eliminated, there is obtained for the determination of the conjugate points the equation

$$\text{II.} \quad \frac{1}{a} + \frac{1}{b} = \frac{1}{f}.$$

which, in its form is exactly the same as that formerly (see Appendix to Chapter IV.) found for the spherical mirror, and expresses distinctly the analogy which exists between mirrors and these lenses.

The equations I. and II., which are primarily deduced for biconvex lenses, hold nevertheless for every form of lens, if we admit the curvature for a plane surface to be indefinitely great ($= \infty$), for a concave surface negative and for a convex surface positive. And according as in the Formula I., the value of f is positive or negative, the lens possesses real or virtual focal points.

CHAPTER VII.

OPTICAL INSTRUMENTS.

46. REFERENCE will here only be made to a few of the numerous applications of lenses to the construction of *optical instruments*.

For experiments in optics intended to be rendered visible to many persons, the light of the sun, on account of its great brilliancy, is employed by preference; un-

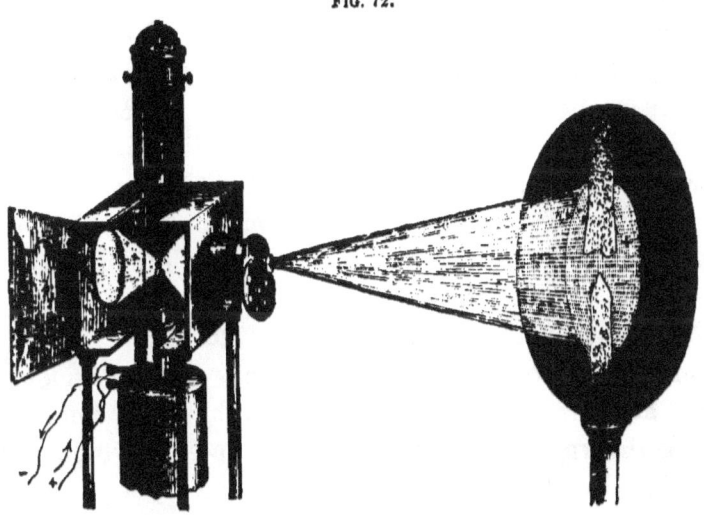

FIG. 72.

Dubosq's lamp.

fortunately, however, in the cloudy northern heavens it is too frequently unavailable, and therefore, in order to be independent of the variations of weather and of daylight, it is customary to substitute for the light of the sun that of an intense artificial light, as for example, that of the electric lamp.

An important, and for many experiments, convenient peculiarity of the rays of the sun is that they are nearly *parallel*. The rays of the electric lamp, on the other hand, issue *divergingly* from the white-hot charcoal points, and hence if they are to be used instead of the sun's rays they must be rendered parallel.

This is effected by means of Dubosq's lamp (fig. 72) which consists of a square box supported on four brass feet, into which the carbon-light regulator (or the lime light, or any other source of light) is introduced.

The light-point is so placed as to be in the focus of a convex lens which is fixed in a moveable frame at the fore-part of the box. By means of the regulating mechanism the carbon points can be made to occupy this position permanently. The rays that fall upon the lens consequently leave the lamp parallel to each other. At the back of the box is a concave mirror, the object of which is to render the rays proceeding in this direction serviceable. For since its centre of curvature is coincident with the carbon points, it returns the rays to their point of origin, from whence they pass to the lens, and having been rendered parallel by this, combine with the rays emanating from the points which are passing directly forwards.

The flame can be so used as to produce a greatly magnified image of the form of the carbon points, and the play of the arc of flame; for if the lens be drawn a little way out of the tube so that the distance of the

OPTICAL INSTRUMENTS. 97

charcoal points is somewhat greater than its focal distance, an inverted and enlarged image of them (fig. 72) is thrown upon the opposite screen. We see between the white-hot carbon points the far less brilliantly luminous violet arc of flame in flickering movement. From time to time white-hot particles are detached from the blunt and excavated positive carbon point, and fly across to the negative point, which remains sharp; small globules are seen moving hither and thither on the surface of the carbon, as though they were in a state of ebullition. These are particles of molten silex which are unfortunately present even in the best carbon points, and by their restless movements occasion the flickering of the luminous arc, whilst, if they happen to occupy the hottest part of the carbon points, they cause an immediate diminution in the intensity of the flame.

47. The experiment just described is identical in principle with the action of the *magic* lantern (fig. 73). It is dependent on the property that convex lenses possess of forming outside of or beyond twice their focal distance an inverted and enlarged image of any object situated on the opposite side between their focal distance and twice their focal distance.

Pictures or photographs serve as objects, and they are placed in a slit in front, *a b*, and are strongly illuminated by the light of the lamp *L*, placed within the box, which is intensified by the lens *m m* and the concave mirror *H H*. In front of the slit is a lens or a combination of two lenses, which act like one of short focal distance, and can be moved by means of a sliding tube. These throw an enlarged image of the object upon the screen. The magic lantern has proved **of** great service in illustrating scientific lectures, in addition to the amusement it affords by its phantasmagoric representations, dissolving views, chromatropes, &c.

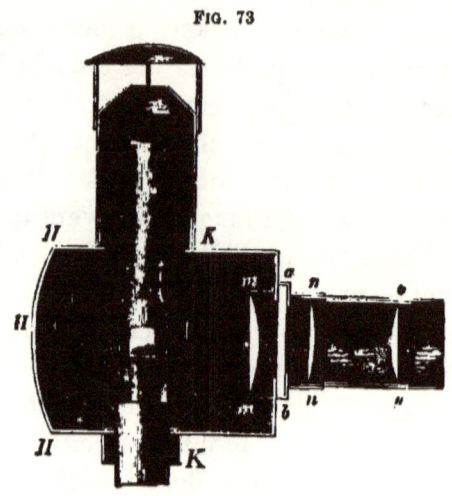

Magic Lantern.

48. The sun or solar microscope (fig. 74) is founded upon the same principle, though it is devoted to thoroughly scientific objects. Its most essential part is a convex lens of short focus, placed in a small tube L, and throwing a greatly enlarged image upon a screen of any small firmly-fixed object, usually between two glass plates, and placed somewhat beyond the focus of the lens L. But since the amount of light proceeding from the small object is diffused over the relatively enormous surface of the image, it is easy to understand that the object must be very brilliantly illuminated if the image is not to be too faint.

The strong illumination of the object is effected by means of a large convex lens placed at the extremity of the wide tube constituting the body of the instrument; this unites the rays of light required for the illumination into its focus.

By means of the screw *C* the object can be placed in this focus, whilst the screw *D* serves to move the lens *L* until the image is thrown with precision on the screen. For the purpose of illumination either the light of the sun may be employed, in which case the

Fig. 74.

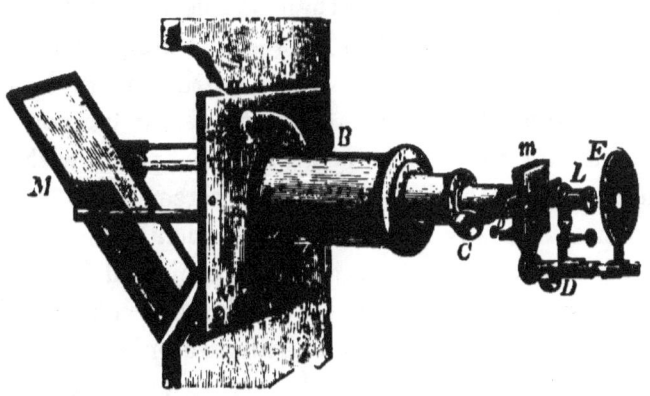

Solar Microscope.

apparatus constitutes, as in our figure, a true 'solar microscope,' which can be placed in the aperture of the Heliostat, or the apparatus may be attached to the frame of a Dubosq's lamp, and the illuminating power obtained from the lime or electric light, in which case the superfluous names of 'photo-electric microscope' and 'oxy-hydrogen microscope' have been applied to it.

The solar microscope proves of great service for the objective representation of small objects in scientific lectures. During the siege of Paris such a microscope, illuminated by the electric light, was made use of in order to project upon a screen and render available for several copyists images of the tiny photographic despatches that were brought by the carrier pigeons.

49. If a convex lens be fitted into the opening of a shutter of a darkened chamber, a variegated picture appears upon the opposite screen, like those which we formerly (§ 13) obtained from a small opening without a lens, but of greater clearness and sharpness. For the lens projects real inverted images of external objects situated at more than double its focal distance upon a screen which lies between its single and double focal distance. But inasmuch as the external objects are situated at very variable distances, it cannot be expected that the images of all should appear with equal sharpness of outline upon the screen. In fact the screen can easily be arranged in such a manner that the image of a distant tower is projected with sharp outlines; but then the leaves of a tree near at hand appear indistinct and confused. In order to obtain a distinct image of the tree the screen must be removed to a somewhat greater distance, but the definition of the outline of the tower is then again sacrificed.

These defects in the definition are nevertheless less considerable than might at first sight appear. It need only be called to mind that if an object is removed from twice the focal distance from the lens to infinity, its image moves over merely the short distance that intervenes between the double and the single focal distance; a great difference in the distance of the object thus corresponds to only a small shifting of the image-plane, and indeed this is smaller in proportion to the remoteness from the lens of the nearest object the image of which is cast upon the screen. It follows that all objects lying beyond certain limits are depicted with tolerably satisfactory sharpness of definition upon a plane situated near the focal point.

The dark chamber the object of which is to keep out collateral light from the image, may be replaced by a box the interior of which has been blackened (fig. 75).

Fig. 75.

Camera obscura.

The lens is fitted into a metal tube i which can be made to slide in the draw tube h by means of a screw the head of which is shown at r. The box a is open at the back and receives a second box b, open in front; in this is a plate of ground glass, the place of which can be shifted by pushing in or out the box b, and which receives the image.

The nearer the object the image of which is cast upon the ground-glass screen is, the further must the box b be withdrawn from the box a. The fine adjustment is effected by the movement of the lens by means of the screw r.

This apparatus, which in its now portable form has received the name of *camera obscura*, remained a mere plaything from the time of its discovery by Porta in the sixteenth century until recently, when its fleeting images have been successfully fixed by photography.

It has now, however, risen to be the chief implement of this highly developed branch of art.

The human eye is only a small camera obscura of wonderfully perfect construction. The crystalline lens, in common with the transparent refracting media filling the globe, casts upon the retina lining its interior an inverted real image of the external world, the impression of which is conveyed to our minds by the functional activity of the optic nerves. The physiological and psychological processes by means of which, in addition to the physical, vision is effected, do not belong to the domain of physical optics. Their consideration, as well as the physiology of the organs of vision, must be passed over.

50. The system of lenses we have here described projects *real* images, which when received upon a screen become apparent to many observers simultaneously. We shall now refer to a series of optical instruments the *virtual* images of which are only visible to a single observer.

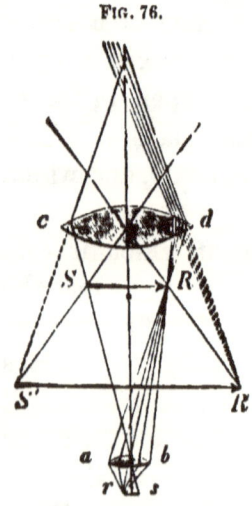

Fig. 76.

Action of the Microscope.

Every instrument by means of which enlarged images of small and near objects are seen, is called a 'microscope.' In this sense the lens above mentioned (p. 87) must be regarded as a '*simple microscope.*' The *compound microscope* possesses a far greater sphere of usefulness. It consists essentially of two convex lenses (fig. 76), which are placed upon a common axis at a distance of somewhat more than the sum of their focal distances. One of these lenses (a b) of very short focus is applied to the object, and is therefore termed the *objective*. It projects to SR an inverted and enlarged real image of any small object (r s), placed at a somewhat greater distance than its focus, which acts as a luminous object to the glass nearest the eye, or *ocular*. This image is seen as the *virtual* image $S' R'$, still further enlarged by means of the ocular, from which it is somewhat less distant than the focal distance of the lens.

Fig. 77 exhibits the form and arrangement of the ordinary microscope. The ocular n, and the objective o, are placed in a vertical tube, which, owing to its being accurately fitted into a brass sheath, n, is moveable with slight friction.

FIG. 77.

Microscope.

The fine adjustment is effected by turning the head of the screw, k. The object, which is usually transparent and fixed upon a glass slide, is placed upon the stage, $p\ p$, and illuminated by light reflected from below by the mirror, s.

If the tube of the microscope be drawn out so far that the image $S R$ is formed outside of or beyond the focal distance of the ocular lens, this lens projects a *real image* of the image $S R$, which can be received upon a screen. In order, however, that this enlarged image should be sufficiently luminous, the small object must be very strongly illuminated by the light of the sun, or by that of the lime light or electric lamp. The light intended for illumination must therefore be concentrated upon the object a by means of a large convex lens, *l*

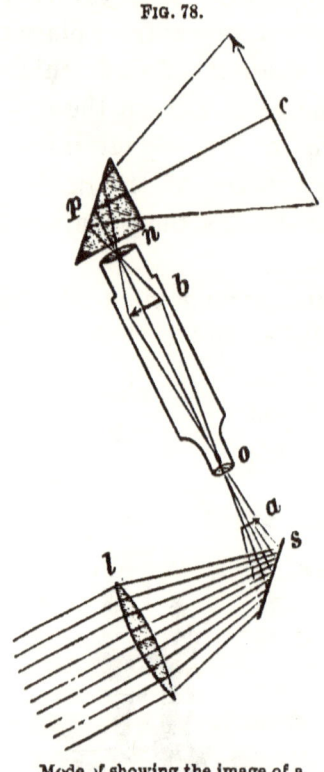

FIG. 78.

Mode of showing the image of a microscope as an object.

(fig. 78), aided by the mirror *s*. The real image of the image *b*, which, on account of the vertical position of the microscope tube, must be formed on the ceiling above, is thrown to the side towards *c* upon a paper screen by means of the prism *p* set at the angle of total reflexion. This arrangement enables us to make use of any ordinary microscope as a solar microscope.

51. The essential features of Kepler's, or the astronomical telescope, (fig. 79) are that two convex lenses, namely, an objective, *o o*, of longer, and an ocular, *v v*, of shorter focus, are placed on an axis common to both at about the distance from each other of the sum of their focal distances. The objective forms near its focus an inverted real image, *b a*, of a remote object, *A B*, which is seen through the ocular, as through an ordinary lens, in the form of an enlarged virtual image, *b' a'*. The visual angle *b' m a'*, under which this image is perceived, is larger than the visual angle, *A C B*, under which the object would be seen by the naked eye, which

explains the magnifying, or, if we may so call it, approximating action of the instrument. As regards the further arrangement of Kepler's telescope, the objective is placed at the anterior extremity, *k*, of a tube of appropriate length (fig. 80), which at the back part is provided with a narrower piece, in which the tube, *t*, containing the ocular, *o*, can be moved to effect perfect definition by means of a screw. Very large instruments of the same kind employed for astronomical observations are called *refractors*.

Kepler's telescope is rendered much more serviceable, not only for astronomical purposes but also for physicists and engineers, by means of the cross threads. These consist of two fine threads of a spider's web, which are arranged at right angles to each other, decussating exactly in the axis of the telescope, and are placed at the point where the image, *b a* (fig. 79) is formed, in consequence of which they must necessarily be seen distinctly with the ocular. If the image of a remote object, as, for example, that of a fixed star, appears at the point of decussation of the threads, the axis of the telescope is directed straight to this point, and its position gives the direction of the visual line from the eye to the star. Kepler's telescope is

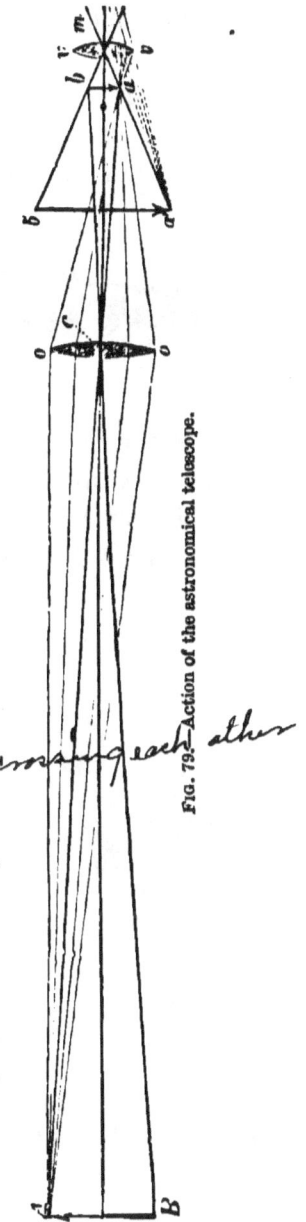

Fig. 79.—Action of the astronomical telescope.

therefore employed in all our instruments for measuring angles.

In the determination of the index of refraction (§ 35), it enables us to measure the slightest deflection effected by a prism. An instrument termed a *theodolite* is made

Astronomical telescope.

use of for the same purpose (fig. 81); it consists of a horizontal disk capable of rotation around its centre (the indicator disk), and a telescope supported upon trunnions. Two markers exactly opposite each other (Nonia) of the revolving disk point to an immoveable circle (limbus) surrounding it, which is divided at its circumference into degrees. In order to determine the deflection of a prism, the telescope is first directed to a narrow and remote source of light; as, for example, a vertical slit in the shutter, until the image of the slit coincides with the vertical thread of the cross threads, and the nonia are read off. The telescope with the indicator circle is then turned till the slit is again perceived to coincide exactly with the cross threads through the prism placed in front of the objective, and the nonia are again read off. The difference between the two readings gives the angle of deflection, $b\ m\ c$, sought for.

Instrument for measuring the prismatic deflection.

In the mirror sextant also (fg. 24), a Kepler's telescope is usual for exact vision.

If the tube containing the ocular of a Kepler's telescope be moved so that the image, $b\,a$ (fig. 83), is more distant from the eye-piece than its focal distance, a real but inverted (and therefore in regard to the object itself erect) image of the image $b\,a$ is projected. In this way an image of the disk of the sun may be thrown upon a screen one mètre in diameter, in which the sun spots are plainly visible.

52. By means of Kepler's telescope objects are seen inverted, which is of little importance in astronomical observations, but is objectionable in the observation of remote objects upon the surface of the earth.

This inconvenience is overcome by replacing the simple ocular acting like a lens by a feebly magnifying compound microscope, which again inverts the inverted image. The compound ocular of the terrestrial tele-

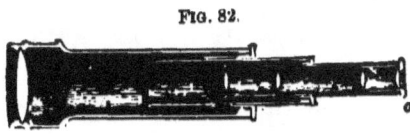

FIG. 82.

Terrestrial telescope.

scope is usually composed of four convex lenses fixed in one tube. This arrangement is seen in fig. 82, which represents a portable telescope with draw tubes, or, in other words, one that is capable of being shut up.

53. Objects are also seen erect with the Galilean, or Dutch telescope. In this form of the instrument the real image, $b\,a$ (fig. 83) of the object $A\,B$ thrown by the convex objective, $o\,o$, is not formed, for the rays here, converging as they do towards every image, strike

108 OPTICS.

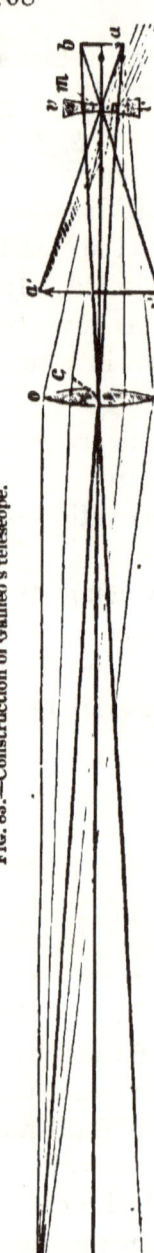

Fig. 83.—Construction of Galileo's telescope.

the concave ocular, $v\,v$, which renders them so far divergent that they appear to come from the vertical erect image $a'\,b'$. Fig. 83 shows very distinctly the course of the rays of light proceeding from the point A of the object. For the image $a'\,b'$ to be seen under a larger visual angle than the object looked at with the naked eye, the virtual focal distance of the ocular must be smaller than the real focal distance of the objective, and the two lenses are accordingly placed at about the difference of these two distances from each other.

The usual form given to the instrument is shown in fig. 84. As no real image is formed by the objective, no cross wires can be inserted; Galileo's telescope is consequently not applicable as a measurer.. Nor again is it possible to obtain any very high magnifying power by its means. On the other hand,

Fig. 84.

Galileo's telescope.

on account of its small length it is extremely convenient as a pocket telescope, and is appropriate therefore for the use of opera glasses (with double or

OPTICAL INSTRUMENTS.   109

triple magnifying power), and to the so-called field glasses, which are able to magnify 20 or 30 diameters.

54. It is very intelligible that on account of the very similar behaviour of lenses and spherical mirrors, telescopes can be constructed in which a concave mirror plays the part of the objective. Fig. 85 shows the construction of a Newtonian telescope. The concave mirror, *S S*, placed at the bottom of a correspondingly

FIG. 85.

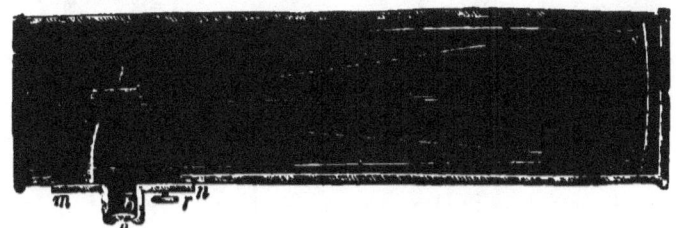

Action of Newton's reflecting telescope.

wide tube, open in front, collects the rays of light coming from a remote object to form a real inverted image at *a*. Before, however, the union is effected they are thrown to one side by a plane mirror, *p*, inclined at an angle of 45° to the axis of the tube, so that the image is thrown to *b*, when it can be observed in the direction *o b* through the convex ocular *o*, as through a microscope.

The reflexion of the small image to the side is necessary, because if the little image *a* be looked for from the front, the head of the observer would obstruct the passage of light to the mirror. In the colossal telescopes (Reflectors) of Herschel and Lord Rosse, the mirrors of which are from 1 to 2 mètres in diameter, the use of such a second mirror, and the consequent loss of light, is avoided by a simple artifice. The concave mirror (fig. 86) is a little inclined to the axis of the tube; consequently, the real image, *a*, comes to lie close to the circumference of the tube, and can be

Fig. 86.

Action of the reflecting telescope with anterior opening.

observed through an ocular, *o*, in the same. The head of the observer is even here, no doubt, partly in front of the aperture of the mirror, but on account of the large size of the latter it is of little importance. Herschel called his instrument 'a front view telescope.'

In using Newton's reflecting telescope the observer has the object looked at to his side; in a front view telescope he turns his back upon it. This circumstance, which excludes direct vision for searching purposes, as well as the inversion of the image, render both instruments inconvenient for the observation of terrestrial objects. In Gregory's reflecting telescope, the external appearance of which is shown in fig. 87, these evils are avoided. The concave mirror, *s s* (fig. 88), is perforated by a circular opening in its centre, and the

Fig. 87.

Gregory's reflecting telescope.

OPTICAL INSTRUMENTS. 111

ocular, o, is placed in a tube behind this aperture. The diminutive inverted real image of a remote object is formed at a, somewhat beyond the focal distance of a small concave mirror, V. This throws to b a once more inverted, and consequently in relation to the object, erect image, which may be looked at through the ocular as with a lens. The fine adjustment is effected by

FIG. 88.

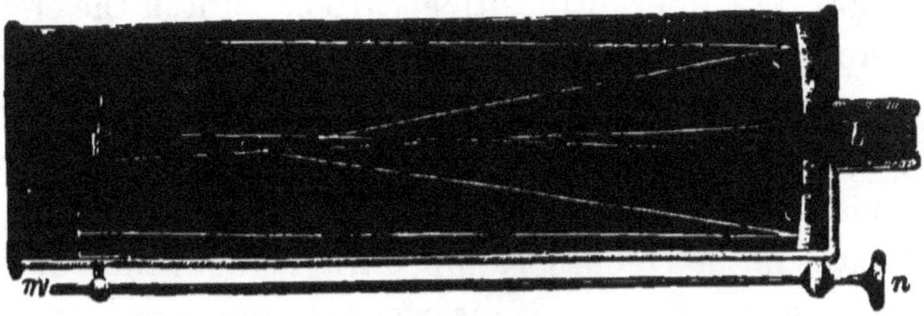

Action of Gregory's reflector.

shifting the little mirror, V, by means of the shaft, $m\ n$, which is provided at m with a screw and at n with a head for turning it. (It is only in the construction of very large instruments that reflectors offer any advantages over refractors.) The use of the smaller reflecting telescopes was formerly very general, when the mode of production of objectives in the perfection desired was not understood; they give, however, only faint images, and cannot now compete with refractors, though very recently they have again undergone great improvement by the application of silvered glass instead of easily oxidisable fused metal mirrors.

CHAPTER VIII.

DISPERSION OF COLOUR.

55. THE inferior (positive) carbon point of the electric lamp is now to be replaced with a thick cylinder of carbon excavated on its free surface for the reception of substances the behaviour of which in the arc of the electric flame is desired to be investigated. After placing the apparatus in the Dubosq's lamp, a fragment of the wax-like, silvery metal Sodium is inserted into the carbon cup, and the two poles are approximated. At the instant of their contact, the current passes through the carbon electrodes and the little ball of metal, which quickly evaporates and fills the arc of flame with its vapour. The whole process may be distinctly followed upon a screen on which an enlarged image of the carbon pole is thrown by the lens when somewhat drawn out. Owing to the metal vapour which rises from the inferior carbon point, the flame acquires a higher degree of conductivity. The poles can therefore be removed to a much greater distance from each other without extinguishing the arc of light which now forms a long flame, shining with a dazzling yellow light (fig. 89), whilst the carbon points, on account of their greater distance from one

FIG. 89.

Vaporisation of metal in the arc of the electric flame.

another, glow much less brightly, and give off much less light than was the case in the experiments (§ 46) formerly made with pure carbon points.

This yellow light of the vapour of Sodium glowing in the electric flame may now be used for other experiments. In the first place the lens of the Dubosq's lamp may again be pushed to a sufficient distance inwards to allow its focus to be situated in the arc of light; its rays are then rendered parallel.

The opening from which a broad cylinder of rays now emanates is closed with a cap having a small vertical slit in it, and the slender beam of parallel rays proceeding from the slit falls upon a convex lens.*

If the lens be placed in a proper position, it throws a well-defined image of the narrow slit upon the screen, which of course exhibits the yellow colour of the source of light employed.

A prism is next placed in the erect position behind the lens in such a manner that its refracting angle is vertical, and is consequently parallel to the slit. The light proceeding from the lens is deflected away from the refracting angle of the prism, and the image of the slit is exhibited, shifted laterally upon the screen, but otherwise unaltered, appearing as a slender vertical yellow streak. (The prism as in all cases is arranged so as to give its minimum refraction.) Up to this point the experiment teaches nothing new. Everything takes place as might be anticipated from our knowledge of the action of lenses and prisms. But if the electric current be interrupted, in order that a new and clean carbon point may be inserted and a frag-

---

* The lens must be *achromatic*. See Chapter IX.

ment of Lithium deposited in its cavity,* the arc of flame assumes a splendid red tint, as does also the image of the slit, whether thrown directly upon the screen or displaced by the prism. We observe, however, that the deflected image is now less distant from the position of the direct image than in the previous experiment. *The red light of Lithium is thus seen to be less strongly refracted through the same prism than the yellow light of Sodium.*

The same experiment may be repeated, taking a fresh piece of carbon each time, with the metals Thallium and Indium. *The splendid green light of Thallium is more strongly refracted than the yellow light of Sodium, whilst the blue light of Indium undergoes a still stronger refraction than that of Thallium.*

It is thus seen that the four kinds of light which have been compared, besides the differences of colour they present to the eye, differ amongst themselves in the circumstance that their refrangibility is progressively greater in the order, red, yellow, green, and blue.

A mixture of the four metals, Lithium, Sodium, Thallium, and Indium may now be placed upon the lower carbon pole. The glowing vapours of all four metals are thus present at the same time in the flame. In the first place, let the direct image of the slit which the lens throws upon the screen without the intervention of the prism be considered. As in the previous experiment, it appears as a bright sharply-defined vertical line, in which nevertheless it is impossible to distinguish any definite tint of colour. The impression received might rather be called that of 'white' light.

* Instead of the metal itself, one of its salts, as for instance the Lithium carbonate, may be used.

On placing the prism again behind the lens, there appear upon the screen no longer one but *four* refracted images of the slit. We see the four coloured bands, which we had before us in the previous experiment

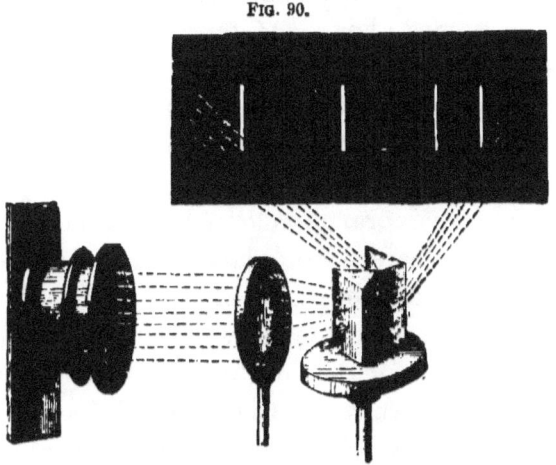

FIG. 90.

Different deflection of different coloured rays of light.

separately, now coincidently one beside the other, each occupying its own proper place,* and each being arranged in order according to its specific refrangibility (fig. 90).

The white light of the electric flame is consequently *compound*, or is a *mixture* of four different kinds of light, which, owing to their different refrangibility, are separated from one another by the prism. Neither of the kinds of light composing the flame undergo any further decomposition by the prism, and hence they are termed *simple* or *homogeneous* light. The prismatic decomposi-

* It is obvious that the prism can only be arranged with precision for the minimum deflection of one kind of light. At the same time if this be the case for one kind of light, as for instance for the Sodium, the refraction of the other kinds of light must be *nearly* at its minimum.

tion of compound light into its constituents, by reason of their different refrangibility, is called the *dispersion of light*.

It is not every chemical substance which, when brought into the electric flame, gives so simple a light as the four named above. If, for example, Strontium, or a salt of this metal, be placed on the lower carbon point, the arc of flame assumes a brilliant red colour, which, however, is not homogeneous like that of Lithium,* since by breaking it up with the prism a group of red and orange-coloured lines may be obtained upon the screen, and lastly, at a considerable distance from them, a beautiful blue line, none of which, however, coincide with the lines of any of the above-mentioned metals, for the brightest red band is somewhat more strongly refracted than the Lithium band, and the blue band is less refracted than the Indium band.

The arc of flame is coloured yellowish green by a salt of Barium. By prismatic dispersion, a group of orange-yellow and green lines are obtained of which again none agrees with those above mentioned in its refrangibility. A characteristic line or group of lines thus corresponds to every metallic element, and serves to indicate its presence in a mixture of luminous vapours.

56. The same method of decomposing light which

* The light of Lithium is, however, itself not completely homogeneous, since in addition to the red, it contains an orange-coloured constituent which is refracted more strongly than the red of the Lithium and yet less strongly than the yellow of Sodium. The Indium further shows besides the blue a still more strongly deflected violet stria. The orange-coloured constituent of the Lithium light as well as the violet of the Indium light being very faint as compared with the red of the former and the blue of the latter, are for the time neglected in the above experiments. The yellow light of Sodium, on the other hand, as well as the green of Thallium, may be regarded as homogeneous kinds of light.

has previously been made use of in examining the light of the electric flame saturated with metallic vapours, may now be applied to the dazzling light of the glowing carbon points itself. For this purpose the earlier arrangement in which both poles consist of small cylinders of carbon may be reverted to. The flame is short between their approximated extremities, and its feeble light is far surpassed by the glow of the white-hot carbon points. Before the prism is interposed, the lens throws upon the screen a sharply-defined *white* image, the slit having a height of about 30 centimetres (13 inches), and very small breadth. If the prism be now placed behind the lens, there appears deflected laterally upon the screen a beautiful coloured band which stretches horizontally to the length of nearly a mètre, but which preserves the height of the slit in the vertical direction (about 30 centimetres). The band shows at the end which lies nearest to the slit a beautiful red, then follow in order the colours orange, yellow, green, light blue, indigo, and finally violet. No one of the colours is sharply defined from the adjoining ones, but each passes into the next through all possible intermediate tints. This coloured band (indicated in fig. 90 by shadow tinting), is called the *Spectrum*.

The experiments made above with the electric light point out how the formation of the spectrum may be explained. Every homogeneous kind of light contained in the beam striking the prism forms on the screen a slender image of the slit exactly at the spot which corresponds to the refrangibility of that kind of light. The spectrum which extends through a wide region of refrangibility is consequently to be explained as the uninterrupted succession of innumerable images of the slit which are arranged in the form of a continuous band. The conclusion is thus arrived at that the white light of the electric glowing carbon is composed of innumerable homogeneous kinds of light, each of which possesses a definite refrangibility in regard to the prism. *The refrangibility continuously increases from the red which is the least, to the violet which is the most, refrangible light.*

That the colours of the spectrum are really homogeneous may be proved by the following experiment. The spectrum is received upon a screen in which is a narrow vertical slit (fig. 91). If this be placed in the middle of the green this coloured light only passes through it, and it undergoes no further decomposition if it be made to pass through a second prism placed behind the slit. Under these circumstances it is merely deflected, without any alteration being effected in its colour, and is consequently demonstrated to be homogeneous. The same holds for all the other colours of the spectrum. The groups of lines produced by the metallic vapours may also be regarded as spectra in which only a limited number of kinds of light (or even only a single kind) is represented. In this sense, for example, it is said that the spectrum of Lithium consists of a red and of an orange red, that of Thallium only of a single green line. In opposition to this interrupted spectrum, that of the carbon points is called an *uninterrupted or continuous spectrum.*

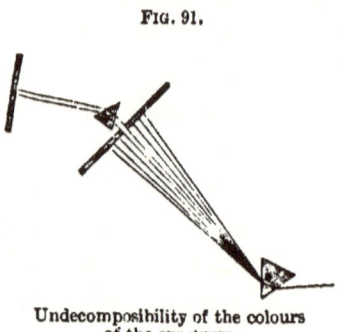

Fig. 91.

Undecomposibility of the colours of the spectrum.

In giving an explanation of the continuous spectrum as a succession of closely-arranged images of the slit, it is requisite to explain why a narrow slit parallel to the refracting angle of the prism is selected as the opening for the incident rays. If the aperture had some other form, as for instance a circular one, the several images refracted through the prism would overlap one another at their edges, as is shown in fig. 92, each colour would mingle with the adjoining one, and no part of the spectrum thus obtained would exhibit a pure and homogeneous colour. By the adoption of a slit placed parallel to the angle of the prism this evil is to a great extent avoided, and in point of fact the spectrum becomes purer and the dispersion into homogeneous colour more complete the narrower the slit is made.

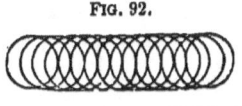

Fig. 92.

Impure spectrum obtained by the use of a circular opening.

57. As white light is a mixture of the various coloured rays of the spectrum, these must conversely be capable of being combined together again to form white light. In fact, if the spectrum be allowed to fall upon the anterior surface of a large lens l (fig. 93),* all the rays

Fig. 93.

Combination of the colours of the spectrum to form white light.

proceeding from a point s of the posterior surface of the prism unite in the conjugate point f, and thus throw upon a paper screen placed at this point an image of

---

* The lens must be *achromatic*.

the posterior surface of the prism in which the dispersed rays reunite. *This image is white.*

It immediately ceases to be white however if one of the colours be abstracted from the mixture. If, for example, the red and orange rays are received on a prism of small refracting angle (fig. 94) placed behind the lens, these are deflected and produce at the side, at *n*, a reddish coloured image. The image *f*, in which still the yellow, green, blue, and violet rays unite, now exhibits a greenish mixed colour. These two reddish and greenish colours must when mingled together (which can be immediately effected by removing the prism *p*) obviously produce white light again, for the one contains exactly those kinds of rays required by the other to form that mixture which we call white. Two colours, which in this way form white by their union, are called *complementary* colours. As the prism is gradually moved along the whole length of the spectrum other colours constantly become deflected to the side, and the images *n* and *f* exhibit successively an entire series of complementary pairs of colours. By this means we learn that red and green, yellow and blue, greenish yellow and violet tints are complementary to one another.

In order to mingle any two simple colours a screen with two vertical slits *a* and *b* is placed before the lens *l* (fig. 95), the distance and breadth of which can

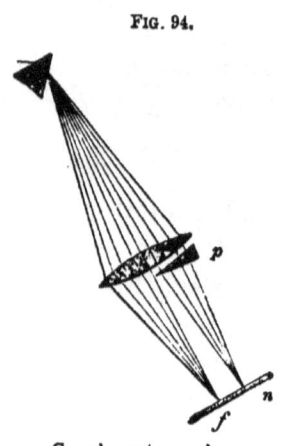

FIG. 94.

Complementary colours.

be altered at pleasure; it follows then that only those parts of the spectrum are combined in the image *f* which have traversed these slits. From red and violet a full purple-red is thus obtained, from blue-violet and orange a delicate rose colour, but out of Indigo blue and yellow — *white*. Thus in order to obtain the impression of white for our eyes, the co-operation of all the colours of the spectrum is by no means necessary, but as Helmholtz first showed, white may be produced by the combination of *only two* homogeneous colours. Amongst the homogeneous colours complementary to each other are red and greenish blue, orange and clear blue, yellow and dark blue, and greenish yellow and violet. It is generally found that for each part of the spectrum from the red end to the beginning of the green, there is a complementary spot in that part of the spectrum which extends from the commencement of the blue to the violet end. The green spectrum colour alone possesses no simple colour, but only a compound one complementary to it, namely, purple.

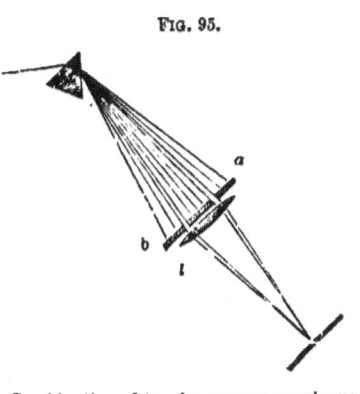

FIG. 95.

Combination of two homogeneous colours.

58. The refraction of compound light is in all instances accompanied by dispersion. If for example a beam of solar rays be allowed to fall upon a prism, this is not merely deflected, but becomes at the same time spread out like a fan, producing upon a screen a solar spectrum which is composed of the same colours in the same

sequence as the spectrum of the glowing electric carbon points.*

The dispersion of the colours of the solar rays is exhibited on the most magnificent scale by Nature herself in the splendid phenomenon of the rainbow. A rainbow is seen whenever the observer turns his back to the unclouded sun and looks towards falling rain.

The following experiment will explain the mode in which the rainbow is formed by refraction and internal reflexion of the solar rays in the spherical rain-drops.

Upon a glass sphere k filled with water and having a diameter of 4 centim. ($1\frac{1}{2}$ in.) a beam of solar light of equal or greater diameter than the sphere is allowed to strike horizontally, and there is then seen, upon a large screen $s\,s$ placed in front of the sphere, and perforated in its centre to allow the passage of the incident rays, arranged concentrically to the aperture and at a distance from it which is nearly equal to that of the sphere from the screen, a beautifully coloured circle, in fact a circular spectrum, the colours of which are arranged concentrically and in such a manner that the red is outside and the violet on the inside. At a still greater distance from the centre of the screen a second similar but much fainter circle is observed, the colours of which however succeed one another in the inverse order, the red appearing on the inside and the violet at the outer periphery.

The first circle is formed by rays which have pene-

---

* If it be required to investigate the phenomena of refraction apart from the influence of dispersion, homogeneous light must be employed. On this ground, in investigating refraction through a prism, the aperture of the Heliostat was formerly (§ 35) closed with a red glass which only permits red and nearly homogeneous light to pass through it.

trated the sphere and have been reflected from its posterior surface, emerging again at its anterior surface.

By reason of this twofold refraction and a single internal reflexion, as is shown in fig. 96, the rays experience a deflection from their original course which differs with the distance of the incident rays from the central ray. By the central ray we mean that which passes through the centre of the sphere; it is reflected upon itself at the posterior surface, and consequently undergoes no refraction. As we pass from this central ray the refraction of the rays begins to increase until at a certain distance it reaches its maximum; from this point onwards to the outermost rays striking the margin of the sphere the amount of refraction again diminishes.

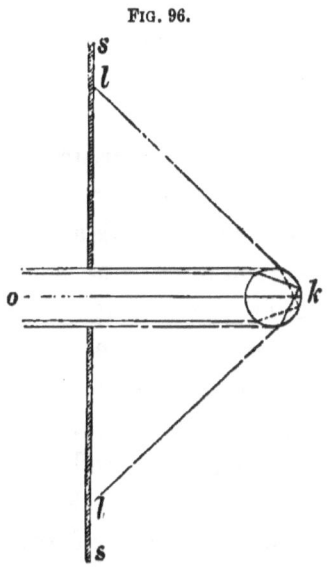

FIG. 96.

Refraction and internal reflexion in a rain-drop.

The most strongly refracted rays which strike the screen at the periphery of the circle cause an illumination that far exceeds that of the single point in the interior of the circular area. If we commence with the rays which undergo the greatest amount of refraction and pass either to the central ray or to the marginal rays, the refraction alters at first very slowly and subsequently very quickly. Consequently the rays which adjoin those that are most refracted associate themselves with the latter after their emergence and augment their light. Those rays, on the other hand, that fall near to one another on other parts of the watery sphere emerge after the second refraction at a distance from each other, and are incapable of producing any well-marked illumination upon the screen.

If the experiment with homogeneous light be repeated, the aperture of the Heliostat being covered with, for example, a red glass, the image upon the screen is reduced to a *feebly illuminated* circular area, which is surrounded by a very bright circular line. The greatest deflection for the red rays amounts to somewhat more than 42° (the angle between ok and kl); the other colours, in consequence of their greater refrangibility, approximate again more to the direction ok of the incident rays, and produce circles the radii of which are successively smaller in the order of their refrangibility. The deflection of the violet rays amounts to about a degree less than that of the red. The direct white light of the sun must therefore produce the circular spectrum which is seen on the screen.

The second iridescent circle is caused by rays which, as is shown in fig. 97, have been twice refracted and twice reflected from within. The least refraction to which such rays are liable amounts to about 51°; for the red rays somewhat less, for the violet somewhat more. This least refraction corresponds to the second circle, the brilliancy of which, on account of the repeated reflexion, is very naturally considerably smaller than that of the former.

FIG. 97. 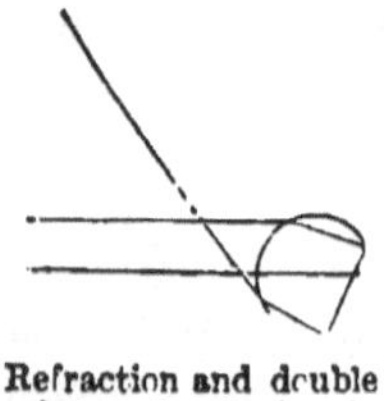
Refraction and double internal reflexion in a rain-drop.

Every falling rain-drop acts in exactly the same manner as the sphere filled with water. An observer

DISPERSION OF COLOUR.

at o (fig. 98), looking at falling rain with his back to a brilliant sun, perceives therefore the light once reflected in the interior of the drops, but only in sufficient strength from such drops as are distant about an angle of 42° from the point of the sky opposite to the sun.* The rays coming from other drops continue their course past the eye unseen. Since the drops $A\,A'$ which remit the red rays toward O are somewhat more distant from the point S than the drops BB', from which the less strongly refracted violet light proceeds, the observer perceives around the central point S the circle described, in which the colours of the spectrum are arranged concentrically from without inwards in the order of their refrangibility. This constitutes the primary rainbow.

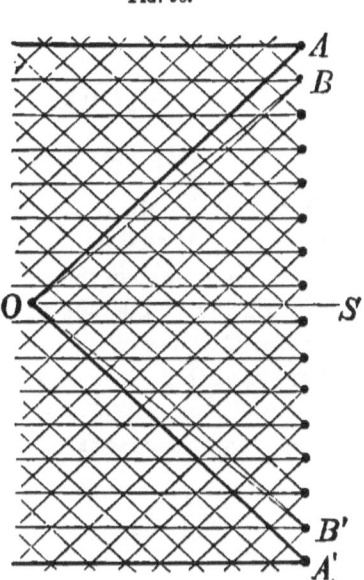

Fig. 98.

Mode of formation of the rainbow.

The much fainter secondary or subsidiary rainbow is distant about an angle of 51° from the point S. It is produced by the rays which, after being twice refracted and twice reflected, have undergone the least possible deflection in the rain-drops; and the reason that the

---

* It is that point where the shadow of the head of the observer would fall if the earth did not hinder it.

colours are arranged in it in inverted order—the red being internal and the violet external—is easy, from what has just been stated, to understand.

APPENDIX TO CHAPTER VIII.

ON THE THEORY OF THE RAINBOW.

IN fig. 99 the circle may represent a sphere of water or a drop of rain. If OS be the straight line drawn from the central point of the drop O to the sun, the line SA parallel to it will represent a ray of the sun striking the drop at the point A. If the radius OA be prolonged to L, the angle LAS or the angle equal to it, AOS, is the angle of incidence (i) of the ray SA. A part, AB, of this ray penetrates the drop under an angle of refraction r, and becomes at B, where it strikes the posterior surface

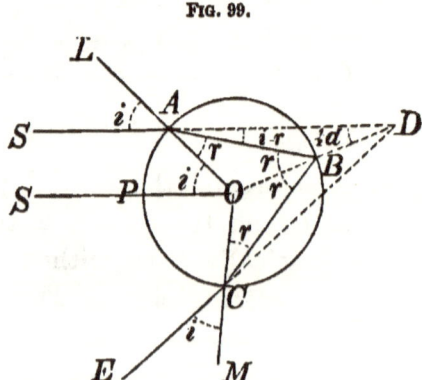

FIG. 99.

Refraction and internal reflexion in a drop of water.

under the angle of incidence $ABO = r$, partially reflected inwards, BC, and returns lastly, after it has suffered some loss by reflexion, inwards at the point C of the anterior surface of the sphere, into the air under the angle of refraction $MCE = i$. Let the ray CE which has been twice refracted and once reflected towards the interior be more particularly considered.

The angle d, which corresponds to the difference between the emerging ray CE and the direct rays from the sun, results in the drawing from the prolongation of the lines SA and CE to their decussation in D. The point D must obviously lie on the prolongation of the radius OB, which divides the whole figure symmetrically, and consequently bisects the angle of refraction d. From the triangle ABD, in which the angle $ADB = \frac{1}{2}d$ and $BAD = i - r$ are opposite to the external angle $ABO = r$, we perceive at once that there is the following relation between the several angles of deflection, incidence, and refraction —

$$\tfrac{1}{2} d + i - r = r ;$$

or, which comes to the same thing,

$$d = 2 (2r - i).$$

This expression shows how the angle of deflection varies with angles of incidence and of refraction; that is to say, with the point where the incident ray strikes the anterior surface of the drop. The median ray SO, for example, which strikes the surface of the sphere perpendicularly at P, is reflected upon itself and undergoes no deflection. The ray CE, on the other hand, which entered the drop at the point A, diverges considerably from its original direction SA. Thus it comes to pass that the innumerable parallel rays that fall upon the upper part PA of the drop, emerge *divergingly* in various directions from its lower part PC. The eye of an observer standing at a great distance and looking towards the lower part, PC, of the sphere, in general therefore receives only a very faint impression of light because almost all the rays proceeding from this point pass by, and only a few reach him.

A stronger impression of light can only be perceived in the event of there being some point upon the anterior surface of the drop in the *vicinity* of which the incident parallel rays are so refracted that after having left the sphere they still *continue their course together* in the direction of their emergence, so that, instead of a single ray, a beam of light composed of a large number of nearly parallel rays reaches the eye, exciting it to a livelier sensation of light.

In order to discover this point, supposing it to exist, let a ray be considered which strikes the sphere very near to the point A. To this the angle of incidence $i + a$ corresponds, which differs only by the very small amount a from that of the ray SA. Coincidently, however, with the angle of incidence the angle of refraction also undergoes a small alteration, β, and becomes $r + \beta$. In consequence of this, the deflection d must also change to a small amount and obtain a new value d'. The relation above found must, however, still always remain between these altered values; that is to say, it must happen that

$$d' = 2(2r + 2\beta - i - a),$$

or that $\quad d' = 2(2r - i) + 2(2\beta - a).$

If this new value of the deflection be now compared with the former one, we perceive that the two values are equal to each other, when

$$a = 2\beta.$$

Hence, in order that two neighbouring incident rays should undergo the same deflection by the drop of water, that is to say, should emerge from it parallel to each other, it is necessary that *the small alteration which the angle of incidence undergoes in passing from one ray to another be twice as great as the corresponding alteration of the angle of refraction.*

Fig. 100 will serve to show how the determination of the position of the point on the periphery of the sphere in which this condition is fulfilled is effected.

The smaller of the two concentric circles represents, as in the preceding figure, the circumference of the drop.

In order to obtain the angle of refraction corresponding to the angle of incidence $AOM = i$, in accordance with what has been already stated respecting the law of refraction,[*] a second circle is to be constructed around the same centre, the radius of which is greater in the proportion of n to 1 (n representing the index of refraction of water). Supposing the radius of the first circle to be unity, that of the second will equal n, and if we now draw through A the straight line QB parallel to OM,

---

[*] See Appendix to Chapter V.

and join the point B where it cuts the circumference of the larger circle with the centre O, BOM will represent the angle of refraction r corresponding to the angle of incidence i.

The segments of the circle MA and MC, which correspond to these angles upon the circumference of the circle having a

Fig. 100.

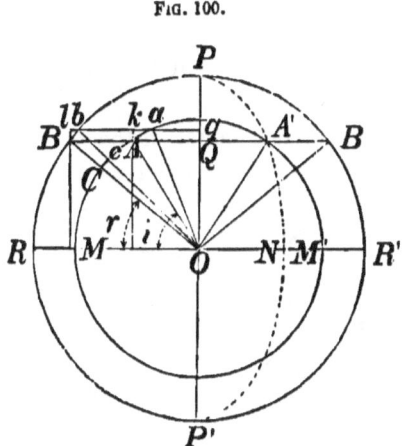

Theory of the rainbow.

radius of 1 may serve as a measure of them. If the same construction be repeated for the larger angle of incidence $aOM = i + a$ around the same segments of the circle $Aa = a$, whilst qb is drawn parallel to OM, we obtain the angle of refraction bOM or cOM, which exceeds the foregoing to the small extent $Cc = \beta$. The arcs Aa and Cc thus represent the corresponding alterations of the angles of incidence and of refraction, and being very small segments of the circumference of the circle, they may, without any very great error, be regarded as rectilinear just as the arc Bb, which corresponds to the small angle of the central point $COc = \beta$ upon the circle having a radius n, and is therefore equal to $n\beta$.

If from the points A and B we let fall the perpendiculars Ak and Bl, and from O the perpendicular Oq upon the straight line qb, we can easily see that the small triangles Aka and Blb are similar to the corresponding and larger triangles AQO and BQO. Hence it follows that

$$\frac{Aa}{AK} = \frac{AO}{AQ}, \text{ and } \frac{Bb}{Bl} = \frac{BO}{BQ}; \text{ or,}$$

if we indicate AQ by v, BQ by ν, the equal segments Ak and Bl by m, and conceive that $AO = 1$, $BO = n$, $Aa = a$, and $Bb = n\beta$; then,

$$\frac{a}{m} = \frac{1}{v}, \text{ and } \frac{n\beta}{m} = \frac{n}{\nu},$$

or also, since in the second equation the factor n appears upon both sides and may therefore be eliminated,

$$\frac{a}{m} = \frac{1}{v}, \text{ and } \frac{\beta}{m} = \frac{1}{\nu}.$$

From these two equations it results that the ratio of the two augments a and β assumes the following form:—

$$\frac{a}{\beta} = \frac{\nu}{v}; \text{ that is to say,}$$

since the coincident changes of the angles of incidence and refraction are constantly to one another as BQ : and AQ; and a is twice as great as β, therefore BQ must be twice as great as AQ, or the point A must bisect the line BQ. In order, consequently, to discover the point A upon the periphery of the sphere of water the neighbourhood of which the parallel rays of the sun are so refracted that they leave the sphere as a parallel beam, the following construction must be applied. Around the circle which represents the circumference of the drop and the radius of which is taken as $= 1$, a second circle is described with the radius n, n being regarded as the index of refraction of water; we now draw the diameter ROR' parallel and the diameter POP' perpendicular to the direction of the incident rays, and amongst the innumerable lines which may be conceived as drawn from the points of the circumference of the second circle parallel to POP' to meet ROR', seek for that one which is bisected by the first circle. The middle point, which must obviously lie in the circumference of the first circle, is the point required. In order to attain this end with certainty, the search must not be entered upon thoughtlessly, but must be proceeded with systematically. If the collective series of lines BQ be conceived to be bisected, innumerable middle points are obtained, amongst which is necessarily the one sought for, which, as a whole, is always a curved line passing through the terminal point P of the second diameter, and through the bisecting point N of the radius OR'. This curved line is obviously an ellipse, the greater semidiameter of which $OP = n$, and the smaller semidiameter $ON = \frac{1}{2} n$. This can be easily constructed, and is seen in the right half of fig. 100.

As the point looked for must lie upon this ellipse as well as upon the circle with the radius 1, it is found immediately as the point of intersection (A') of these two curved lines. The angle of incidence sought for $A'OM' = i$, as well as the corresponding angle of refraction $B'OR' = r$, may now be obtained either directly from the figure by measurement, or more exactly by calculation.

If it be admitted for the sake of argument that the sun emits only the simple yellow light of Sodium, the index of refraction of water for this kind of light is exactly $\frac{4}{3}$. If this value be taken as a base for the construction, we find $i_1 = 59°\ 24'$, $r_1 = 40°\ 12'$,* and since d_1 is equal to $2(2r_1 - i_1)$ the corresponding deflection is

$$d_1 = 42°.$$

In this direction only does a beam of nearly *parallel* rays emerge from the drop, which, because they remain together in the long path to the eye, penetrate it together, and hence occasion a lively sensation of light.

These rays, which emerge parallel to each other from the drop, are distinguished from the rest in another point of view. Their deflection is the *maximum* which the sphere of water is capable of producing on rays of a definite refrangibility. We can easily convince ourselves of this by the following consideration. At the point A, which corresponds to the angle of inci-

* It is remarkable that for the index of refraction $\frac{4}{3}$ the angle of incidence and triple the angle of refraction together from two right angles, that is to say, $i_1 + 3r_1 = 180°$.

dence i_1, as we have seen, the alteration a of the angle of incidence is equal to twice the alteration of the angle of refraction or to 2β. On the other side of the point A, with the greater angle of incidence $i_1 + a'$, to which also a greater angle of refraction $r_1 + \beta'$ corresponds, a' is greater than 2β, because the same also BQ (fig. 100) is greater than $2AQ$. The deflection of this ray is consequently

$$d' = 2(2r_1 + 2\beta' - i_1 - a')$$

or,

$$d' = d_1 + 4\beta' - 2a'.$$

Since a' is greater than $2\beta'$, and therefore also $2a'$ is greater than $4\beta'$ we have, in order to obtain d', to subtract more than to add, consequently d' is smaller than d_1. On this side of the point A, the angle of incidence is smaller than i_1, it is $i_1 - a''$ and the corresponding angle of refraction $r_1 - \beta''$. The deflection d'' which this ray experiences is therefore

$$d'' = 2(2r_1 - 2\beta'' - i_1 + a'')$$

or,

$$d'' = d_1 - 4\beta'' + 2a''.$$

But since because BQ is here less than $2AQ$, a'' is also less than $2\beta''$, we must subtract a greater amount than we add, and d'' is thus less than d_1. The deflection d_1 which the parallel rays experience on their emergence, is thus in fact the maximum which can occur with single internal reflexion.

In fig. 100 the determination of the point A is only effected for the single ratio of refraction $\frac{4}{3}$; for every other index of refraction we must construct according to the same rules another external circle and another ellipse, and thus convince ourselves that the less refrangible rays experience a greater refraction $(= 42° 13')$, and the more refrangible violet rays a less deflection $(= 41° 14')$.

The evidence above adduced constitutes the basis on which the explanation of the primary rainbow is founded.

In regard to the secondary, a brief explanation, after what has just been said, is all that is necessary. Since the deflection which a ray of light has experienced after double internal reflexion is expressed by

$$d = 180° - 2(3r - i)$$

the condition $a = 3{,}3$ must be present for parallel emerging rays.

We find therefore the point of incidence which satisfies this condition if we construct an ellipse in fig. 100, of which the greater axis likewise $= n$, but the smaller axis $= \frac{1}{3} n$. By a quite similar train of reasoning it may then easily be shown that the deflection ($= 51°$ for $n = \frac{4}{3}$) which corresponds to this point is the *minimum* which can occur with double internal reflexion.

CHAPTER IX.

ACHROMATISM.

59. A PURE spectrum of solar light is obtained by allowing it to pass through the vertical slit of the Heliostat, and arranging the lens, prism, and screen as before. At first sight the solar spectrum does not appear to differ from that of the electric light; the succession and division of the colours, the degree of refraction and length of bands of colour is* the same in both cases. On closer inspection, however, of the brightly illuminated surface, we perceive a great number of dark lines, which are disposed perpendicularly to the long axis of the spectrum, and consequently parallel with the slit. These dark lines, the majority of which are extremely fine, though some are very well marked, were first observed by Wollaston (1802), and were subsequently more exactly investigated by Fraunhofer (1814). The last-named observer, from whom they have received the name of Fraunhofer's Lines, distinguished eight prominent lines by the letters A to H. The line A lies at the extremity of the dark red; B and C in the middle of the red; D between the orange and yellow; E in the green; F in the intermediate colour between green and blue; G in the dark blue, and H towards the end of the violet (see fig. 106).

* For the same prisms.

The spectrum of solar light is consequently not continuous, like that of white-hot charcoal, but there are small interspaces which appear to us as fine dark lines. From the presence of these spaces we must conclude that the homogeneous kinds of light corresponding to them are deficient in the light of the sun.

The lines of Fraunhofer constitute well-defined marks, within the gradual transitions of colour of the spectrum which always correspond to the same homogeneous kinds of light, and afford us the means of defining each part of the spectrum, and of discovering it again at all times with certainty. How very useful these points are in our enquiries will be seen as we proceed.

60. Up to the present time a prism of flint glass has always been used for the production of the spectrum. But, in order to compare the dispersion of colour of various substances, three prisms must successively be taken, each of which possesses a refracting angle of 60°, namely, one of flint glass, one of crown glass, and finally, a hollow prism filled with water. The first thing that is observed is that the spectra which they throw are refracted laterally to different extents. That caused by the flint prism is deflected to the greatest degree, that by the crown glass to a less extent, and that by the water prism least strongly. The spectra vary also considerably in length; the spectrum thrown by the flint glass is nearly double as long as that thrown by the water prism.

We may now ask: Is the stronger dispersion of colour exhibited by the flint-glass spectrum simply the consequence of its greater refracting power, or does the flint glass, in virtue of its material qualities, possess a greater power of dispersion than the other two substances? In order to answer this question, we must compare the lengths of the spectra of equal refraction with one another. A flint-glass prism may easily be prepared which shall cause the same refraction in any particular homogeneous kind of light, as, for example, in the rays which correspond to Fraunhofer's line *D*, as a prism of crown glass of 60°. Such a prism of flint glass must obviously have a refracting angle of less than 60°, and one in fact that amounts to about 52°. The crown-glass prism of 60°, and the flint-glass prism of 52°, give spectra in which the line *D* undergoes the *same amount of deflection*. *Notwithstanding this, the flint spectrum from B to H is nearly double as long as that of the crown glass.* From this it may be concluded that the power of dispersion of the flint glass is almost double (speaking exactly, 1·7 times) as great as that of crown glass.

Two similar prisms made of the same material (for example, two prisms of 60° composed of crown glass) of

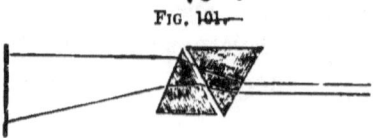

Fig. 102.

Combination of two similar prisms without deflection and without dispersion.

course produce equal refraction and equal dispersion of colour, that is to say, equal length of the spectrum. If they be placed, as in fig. 101, behind one another with their refracting angles in opposite directions, the second one restores to the original condition the refraction as well as the dispersion of colour caused by the first. The white beam of light which penetrates

ACHROMATISM.

the first emerges from the second as white light again, coursing parallel to its original direction, and producing a white image of the slit upon the screen. The combination of the two prisms acts like a thick plate of glass with parallel surfaces, which causes neither refraction nor dispersion. What will occur, we may now ask, if a crown-glass prism of 60° be placed behind a flint-glass prism of 52° with the refracting angle reversed? The deflection of the Fraunhofer's line D disappears; but since it causes nearly twice as long a spectrum as the crown-glass prism, the dispersion of colour is *not* removed, but becomes reversed. We perceive therefore upon the screen in the direction of the direct rays a spectrum of about the same length as that caused by the crown-glass prism, but with the succession of colours inverted.

In making observations upon the spectrum formed by a prism, it is frequently inconvenient that the spectrum should be deflected so far to one side.

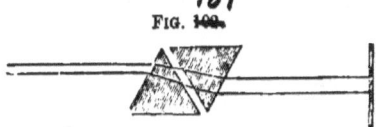

Fig. 101.

Combination of a crown and of a flint-glass prism causing dispersion but no deflection.

The experiment just made, however, shows how the spectrum may be obtained in the direction of the incident rays, and to avoid the necessity of putting the prisms into position on every occasion, they may be cemented together by a transparent substance (Canada balsam). Such a combination is called a *direct vision prism*. Such combinations of prisms are usually made up of three (fig. 103) or of five (fig. 104) prisms; one flint and two crown, or two flint and three crown.

Now a prism of flint glass which throws just as long a spectrum as a prism of crown glass must have its

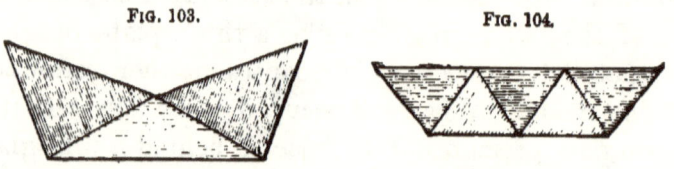

Fig. 103. Fig. 104.

Showing combinations of prisms which cause no deflection (*à vision directe*).

refracting angle about half the size of that of the latter. It causes, however, considerably less deflection. If we combine therefore two such prisms (a crown-glass prism of about 60° and a flint-glass prism of about 30°) placing them in opposite positions (fig. 105), the second abolishes the dispersion of colour

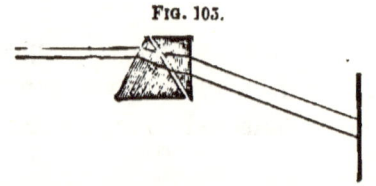

Fig. 105.

Combination of a crown and flint-glass prism, with deflection but without refraction (an achromatic prism).

produced by the first. On the other hand, it diminishes but does not completely remove the deflection. We obtain therefore upon the screen *a white image of the slit deflected to one side*. In the combination of the two prisms we thus possess *a prism causing no dispersion of colour, or an achromatic prism*.

Thus it appears that one of the two actions of a prism, deflection and dispersion, can be abolished without interference with the other, nevertheless only by a combination of at least two prisms made of *different materials*. Two prisms made of the same kind of glass

ACHROMATISM. 139

either abolish both actions simultaneously (fig. 101), or leave both intact.

61. The different power of dispersion possessed by various substances shows that an influence is exerted by the material of which the prism is composed upon the light traversing it. This action may be still further followed if spectra of equal length from *B* to *H* (fig. 106) of a crown-glass prism of 60°, and a flint-glass prism of

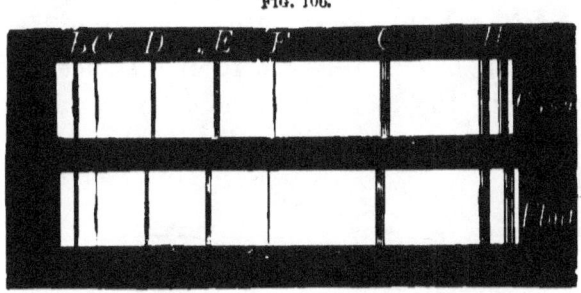

Fig. 106.

Spectrum thrown by crown glass and by flint glass.

30°, be compared, for which purpose the lines of Fraunhofer, which always correspond to the same homogeneous tints of colour, serve as excellent guides. By their position in the two spectra it is rendered evident that the less refrangible rays are more closely approximated in passing through the flint glass, whilst the more refrangible are separated further from one another than by the crown glass; so that although the total dispersion of the two prisms (that is to say, the length of their spectra between *B* and *H*) is exactly the same, their dispersion is different. If, therefore, as previously pointed out, they be added together, the second cannot *completely* abolish the dispersion of the former, and the combined prism is not completely achromatic. The very small dispersion of colour that still remains can only be removed by a properly selected thicker prism, composed again of a third substance. In the meantime, however, it is so small that it may be usually neglected.

62. The laws of light in regard to lenses, of which a knowledge has already been acquired, are only strictly accurate under the presumption that we are dealing with homogeneous light; as, for example, with the light of the Sodium flame. In consequence of the unequal refrangibility of the different coloured rays, an ordinary lens has a different focal distance for each kind of light—the focus of the violet rays (v, fig. 107) being nearer to the lens than that of the red rays (r).

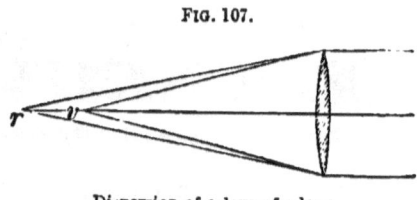

Fig. 107.

Dispersion of colour of a lens.

It is impossible for the rays emanating from a luminous point of white or parti-coloured light to be reunited again into one point; the images thereon are therefore not sharply defined, but surrounded by faint coloured rings. A telescope or microscope with such a lens as an objective would, on account of the indistinctness of its images, be almost valueless.*

The prevention of the dispersion of lenses is always therefore an object of solicitude in practical optics; and before the solution of the problem was discovered by

* We can, however, obtain well-defined images with a microscope thus dispersing light, if we illuminate the object with homogeneous light, such for instance as that of the Sodium flame.

Hall in 1733, and by Dollond in 1757, it was impossible to construct serviceable telescopes, and it was found necessary to take refuge in the less powerfully luminous reflecting telescopes.

That a *single* lens can never be free from dispersion is obvious; but, on the other hand, it is possible to combine two lenses of such nature that each is capable of mutually compensating for or destroying the dispersion of the other. A method by which the desired result may be obtained is indicated by the production of the achromatic prism.

In order to remove, namely, the dispersion of colour of a lens, we place a second lens of opposite action immediately behind it which possesses the same dispersion of colour but causes a different amount of refraction; that is to say, has another focal distance.

We add, for example, to a convex crown-glass lens a concave flint-glass lens; and in order that both should effect equal but opposite dispersion of colour, the virtual focal distance of the latter must be about twice as great as the real focal distance of the former. Their combination then gives an achromatic lens (fig. 108), which unites all the rays emitted from a white point into a white image-point again.

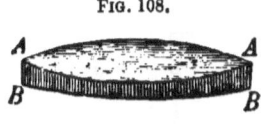

Fig. 108.

Achromatic lens.

For the reason formerly mentioned in speaking of the achromatic prism, we do not even here obtain entire freedom from colour. The amount still remaining is, however, extremely small.

63. The first compound achromatic lenses constructed on this principle were discovered by experi-

ment. The greatest perfection can, however, only be obtained if, instead of the uncertain method of trial, direct calculation be made of the most favourable form for both the flint and crown glass. In order to do this, however, an exact knowledge of the indices of refraction of the kinds of glass for the various homogeneous rays of light is required. The indices of refraction in regard to the red, yellow, green, and other rays, were laid down long ago, but on account of the gradual transition of the rays into each other rendering a sharp definition of their limits impracticable, the numbers discovered were inexact. But when Fraunhofer employed the dark lines named after him as fixed points, he was able to measure exactly the indices of refraction for determinate homogeneous rays, and, proceeding on this information, to construct achromatic objectives for telescopes that have not hitherto been surpassed in the perfection of their performance. The method we have hitherto pursued in order to throw the spectrum as an *object* upon a screen is excellently adapted to exhibit a large number of its peculiarities. If, however, it be desired to make a special study of its characters, and to make measurements, the direct method of observation applied by Fraunhofer has the advantage.

FIG. 109.
Measurement of refraction as practised by Fraunhofer.

In this method a telescope (fig. 109) is placed immediately behind the prism, the objective lens of which, whilst it receives the rays emerging from the prism, throws a spectrum near its focus, which is then seen with the ocular as through a lens. The Fraunhofer lines can thus be seen with extraordinary definition and clearness. The direct method of observation through a telescope also has the advantage that it does not require nearly so much light as the projected image method.

If a divided circle be combined with the observing telescope (fig. 109), we are able, by directing the cross threads successively to each Fraunhofer's line, to measure accurately the slightest differences in their position, and then in accordance with the method above given to determine the corresponding index of refraction. The indices of refraction of some of the more important substances for the principal Fraunhofer lines as thus obtained are given in the accompanying little Table:—

	B	C	D	E	F	G	H
Water	1·3309	1·3317	1·3336	1·3359	1·3378	1·3413	1·3442
Alcohol	1·3628	1·3633	1·3654	1·3675	1·3696	1·3733	1·3761
Carbon bisulphide	1·6182	1·6219	1·6308	1·6438	1·6555	1·6799	1·7019
Crown glass, No. 9	1·5258	1·5268	1·5296	1·5330	1·5361	1·5417	1·5466
Flint glass, No. 13	1·6277	1·6297	1·6350	1·6240	1·6483	1·6603	1·6711
Flint glass of Merz	1·7218	1·7245	1·7321	1·7425	1·7521	1·7725	1·7895

As each substance has a special index of refraction for each kind of ray, it is necessary to point out in every statement respecting an index of refraction, which homogeneous ray is meant, and when, as in the indices of refraction given at p. 60, such a precise statement is neglected, the observation is only approximate, and refers to the middle rays between D and E.

Any Theodolite may be used for the measurement, upon Fraunhofer's plan, of prismatic deflection, and in order that the prism should follow the rotation of the telescope, it must be placed upon a small table attached to the objective end of the telescope. The refracting angle of the prism, which must be known for the calculation of the index of refraction, is determined by means of the reflecting goniometer, p. 34.

64. The determination of the index of refraction can be much more conveniently effected by means of *Meyerstein's Spectrometer*, a representation of which is given in fig. 110. The observing telescope is here directed to the centre of the horizontal divided circle, and is sup-

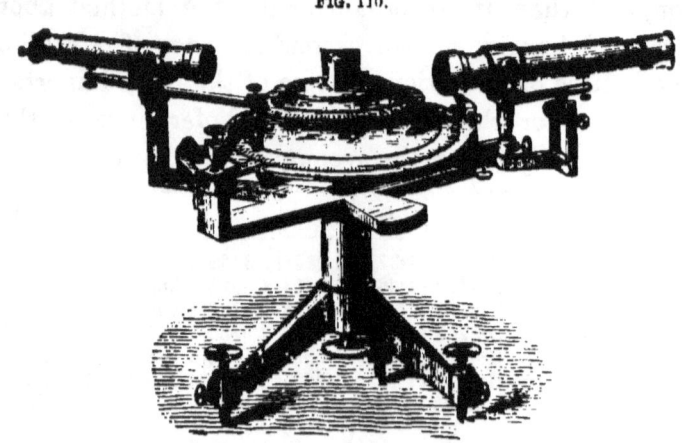

Fig. 110.

Spectrometer.

ported on horizontal arms connected with the vertical axis of the divided circle. This axis rotates in the bore of a metal column supported by three screws giving off above, three horizontal arms. Two of these, which are opposite to each other, carry the indicators (nonia) by means of which the rotation of the divided circle is read off; the third arm carries a telescope directed towards the centre from which the ocular has been removed, and is replaced by a vertical slit. This slit is situated in the focus of the objective lens, so that the rays pro-

ACHROMATISM.

ceeding from it strike the prism as a parallel beam, and traverse it at right angles to its refracting edge, that is to say, each passes through a principal section. Were this condition not fulfilled, the prism would produce, in consequence of the rays directed obliquely to its principal section, a confusion of the image of the slit which would make itself disturbingly perceptible in the spectrum as a curvature of the Fraunhofer's lines. Whilst by means of the 'slit-tube,' the slit can be withdrawn to any distance, it confers upon the Spectrometer the advantage of being applicable to the investigation of the weaker lights.

To obtain parallel rays when employing the method of Fraunhofer, the distance of the Theodolite from the slit must be increased as far as possible; on this account it is especially adapted for sunlight, for when the distance is considerable the feebleness of artificial sources of light is not sufficient; with the Spectrometer, on the other hand, the source of light can be brought immediately in front of the slit, and consequently weaker sources of light can be made the subject of experiment. When the observing tube and the slit tube have exactly the same direction, the slit is seen at the decussation of the threads of the former, and the indicator points to zero upon the divided circle.

We now place the prism (or rather the small tablet supported by three screws on which it stands) in the middle of the instrument, upon a second smaller horizontal divided circle, the vertical axis of which turns in a socket formed by a bore in the axis of the greater circle. We must now turn the observing tube, and with it the great circle, to one side, in o der to perceive the deflected image of the slit, or rather its spectrum;

by turning the small circle the prism can easily be brought into the position of smallest deflection, the amount of which can be read off after accurate focussing by the indicator of the large divided circle.

The smaller divided circle has still, however, a second important use. It forms, if we allow the greater circle to remain fixed, with the slit and observing tube together, a Reflecting-goniometer (p. 34). We can therefore with all necessary exactitude determine by means of this instrument, the Spectrometer, the two qualities which are required for the calculation of the index of refraction, namely, the smallest deflection and the refracting angle of a prism.

APPENDIX TO CHAPTER IX.

ACHROMATIC LENSES.

WHEN two thin lenses are placed one immediately behind the other, as in fig. 108, the deflection which they produce in a point at any distance k from the common axis is equal to the sum of the deflections which each of the lenses would have itself effected. If F therefore indicates the focal distance of the compound lens f, that of the first, and ϕ that of the second lens,

$$\frac{k}{F} = \frac{k}{f} + \frac{k}{\phi}, \text{ or } \frac{1}{F} = \frac{1}{f} + \frac{1}{\phi}.$$

The focal distances f and ϕ of the two separate lenses are, however, different for different coloured rays, for we obtain (according to Equation I. p. 92, for example), the focal distance for red rays

$$\frac{1}{f_r} = \left(\frac{n'}{r} - 1\right)\left(\frac{1}{r_1} + \frac{1}{r_2}\right)$$

for violet, on the other hand,

$$\frac{1}{f_v} = \left(n'_v - 1\right)\left(\frac{1}{r_1} + \frac{1}{r_2}\right),$$

where n'_r and n'_v indicate the indices of refraction of crown glass for red and violet rays, and r_1 and r_2 the radii of curvature of crown-glass lenses.

In the same way we have

$$\frac{1}{\phi_r} = \left(n''_r - 1\right)\left(\frac{1}{\rho_1} + \frac{1}{\rho_2}\right),$$

and

$$\frac{1}{\phi_v} = \left(n''_v - 1\right)\left(\frac{1}{\rho_1} + \frac{1}{\rho_2}\right),$$

where the corresponding quantities for flint-glass lenses are indicated by n''_r and n''_v, ρ_1 and ρ_2. If the combination of the two lenses for red and violet possess the same focal distance the two lenses must be such that

$$\frac{1}{f_r} + \frac{1}{\phi_r} = \frac{1}{f_v} + \frac{1}{\phi_v}.$$

With the aid of this equation and the expressions above given for the several focal distances, the radii of curvature which must be given to the two lenses in order to obtain an achromatic system may be calculated with facility.

CHAPTER X.

SPECTRUM ANALYSIS.

65. If instead of the measurement of indices of refraction the observation and comparison of the spectra proceeding from various sources of light be the subject

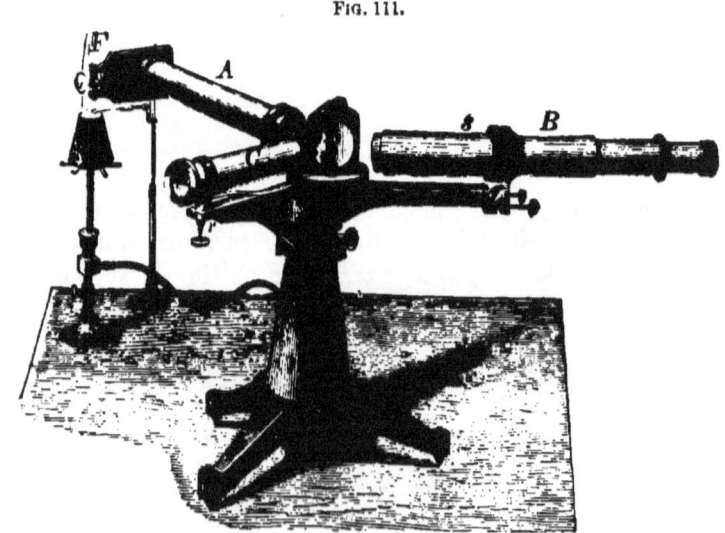

Fig. 111.

Bunsen's spectroscope.

of enquiry, the divided circle of the spectrometer may be dispensed with; and the instrument thus simplified constitutes *Bunsen's Spectroscope* (fig. 111), in which the slit tube A, the prism P, and the observing tube B, are all arranged just as in the spectrometer. In order, however, to obtain the means of measurement within the limits of the spectrum without a divided circle, a very ingenious apparatus has been introduced. A third tube, C (the scale-tube), has at its outer end, at s, a small photographed scale with transparent divisions, whilst at the inner end is a lens which is placed at about its focal distance from the scale. The scale is illuminated by means of a lamp or candle. The scale-tube is so placed that the rays of light that proceed from the scale and emerge parallel to the axis of the tube are reflected at the anterior surface of the prism in the direction of the observing tube. The observer looking into the telescope sees therefore coincidently with the spectrum of the light F, the image of the scale, which may be used as a measure.

As the rays are deflected from their original direction by the prism, the observing tube in the spectroscope just described must be so placed in regard to the slit-tube as to form an angle which is about equal to the smallest deflection of the middle rays. The source of light to be investigated cannot therefore be looked at directly, a circumstance which renders the arrangement of the instrument difficult and its management somewhat awkward. The *direct vision* or *rectilinear spectroscope* (*à vision directe*) which instead of a single prism contains a combination of prisms, so that there is no deflection, is free from this inconvenience (fig. 104). To this class belongs *Hoffman's Spectroscope*, and the little (only 3½ in. long) pocket *Spectroscope of Browning*.

66. By means of the spectroscope the spectra of the glowing vapours formerly thrown upon the screen can be very conveniently observed (objectively). But

whilst for those researches the dazzling light of the electric flame was requisite, the flame of a Bunsen's burner is now sufficient, at least for the light metals (fig. 1). Instead of the metal itself, some of its chemical combinations, or so-called salts, are usually employed. A small quantity of such a salt is melted at the extremity of a fine platinum wire, and introduced into the external hottest part of the feebly luminous flame. The salt is decomposed by the heat; the flame is saturated with the vapour of the metal now set free, and is tinted with a colour characteristic of the metal. With a little Sodium chloride (common salt), for example, we obtain the homogeneous yellow light of Sodium; salts of Lithium and Strontium colour the flame of a carmine red tint; salts of Potassium clear violet; salts of Barium green; and salts of Calcium yellowish red. Analysts had no doubt long employed these characters to demonstrate the presence of the above metals, but the colour of the flame continued to be an uncertain means of recognition until prismatic decomposition was applied as a means of investigation. It was almost impossible, for example, with the naked eye to distinguish between the red flame of Lithium and that produced by Strontium, but if the two are looked at through the spectroscope they exhibit perfectly distinct spectra, which are exhibited on the Spectrum plate (*see* Frontispiece, Nos. 6 and 8). If, again, a specimen of Sodium salt with which only a trace of Lithium is mingled be examined, the presence of the latter cannot be recognised with the naked eye, because its feeble red stain is completely overpowered and concealed by the brilliant yellow of the Sodium. The spectroscope, however, shows distinctly the red line of Lithium close to the yellow Sodium line, each in its place, thus disclosing the chemical composition of the substance in question.

This qualitative method of chemical analysis is termed *spectrum analysis,* and although the spectra of some coloured flames had been known for some time, and their applicability as chemical tests recognised, Bunsen and Kirchhoff were the first who laid down the scientific grounds on which alone a method of investigation could be raised, and who must therefore be regarded as the true discoverers of spectrum analysis. Bunsen and Kirchhoff showed first that the positions which the bright lines of the spectrum occupy are independent of the temperature of the flame; in fact, that the same red colour is obtained and the same two lines, a red and a reddish yellow, are seen in the spectroscope whether the Lithium chloride be volatilised in the flame of a Bunsen's burner or in the much hotter flame of the oxyhydrogen blowpipe. It is to be noted that the brilliancy of the several lines increases with increasing temperature, and thus it may happen that by means of intense heat lines come into view which at lower temperatures are too feeble to be perceived. If, for example, Lithium be volatilised in the electric flame, a blue line is visible in its spectrum, which occupies exactly the same position as the blue line of Strontium. In the flame of the Bunsen's burner it exhibits only the two above-named lines. Moreover, the two observers just mentioned demonstrated that different combinations of the same metals give invariably the same spectrum, whence the conclusion is irresistible that the lines seen in any instance may be regarded as positive evidence of the actual presence of the metals in question.

The spectrum method of analysis is distinguished from ordinary chemical methods by its extreme delicacy. The three-millionth part of a milligramme of a salt of Sodium, an imperceptible particle of dust to the naked eye, is yet capable of colouring the flame yellow and of giving the yellow line of Sodium in the spectroscope. More than two thirds of the surface of the earth are covered by sea, which contains Sodium chloride, or common salt. When waves are raised by the storm and their foaming summits are carried away, fine particles of salt are mingled with the air and carried far over the land; common salt is consequently distributed through the whole atmosphere in the form of a fine dust. On account of this almost constant presence of Sodium chloride, it is scarcely possible to obtain a flame which does *not* exhibit the yellow line of Sodium. It is only necessary to strike a handkerchief upon the table, or to close a book sharply, to make the dust which escapes colour the adjoining Bunsen's flame yellow, and to make the Sodium line appear in the spectroscope. Moreover, in the representation of the spectra of different metals by means of the electric lamp, they can never, as has been seen, be obtained completely free from the Sodium line.

The extraordinary sensitiveness of the spectrum method of analysis led its celebrated discoverers, Bunsen and Kirchhoff, to the discovery of two new alkaline metals that had previously escaped the notice of chemists, Cæsium and Rubidium, the compounds of which occur only in very small quantities in minerals and mineral waters. These spectra are represented in Nos. 2 and 3 of the Spectrum plate. Subsequently, Crookes, from the spectroscopic examination of the crust formed in the lead chambers of a sulphuric acid manufactory, discovered the lead-like metal Thallium (No. 10), and Reich and Richter also discovered, by means of spectrum analysis, in certain ores of zinc the zinc-like metal Indium.

67. The spectrum method of analysis just described has been chiefly applied to the recognition of the alkalies and alkaline earths, for the heat of a Bunsen's burner is insufficient to volatilise the heavy metals and obtain their vapour in a glowing state. To effect this we must seek other means, and we possess them in the electric lamp, which may be used in order to exhibit the spectra of several of the heavy metals upon a screen. If a fragment of zinc be volatilised between the carbon poles a series of beautifully coloured striæ are seen, especially one red and several blue. If now a fragment of brass, which is composed of zinc and copper, be added, in addition to the zinc lines the group of green lines peculiar to copper are immediately observed. By the addition of a little silver the spectrum of this metal appears, which also exhibits a group of green lines, but these are easily distinguishable by their position from those of the copper. It is observable that the inevitable Sodium line is a constant accompaniment of all these experiments.

As a powerful galvanic battery is required for the production of the electric arc of light, spectrum analysis in its application to the discovery of the heavy metals would prove very troublesome were there no more convenient means of converting the metals into luminous vapours. For the purposes of subjective observation through the spectroscope the ordinary electric spark is sufficient, or still better the spark of a powerful induction apparatus, which by means of a few galvanic cells can be maintained in unbroken activity. This apparatus is exhibited in fig. 112, but a detailed description of its construction and mode of action would here be out of place. It is enough to say that if the conducting wires of the galvanic battery are fastened down by the binding screws C and D, electric sparks succeed each other in rapid succession between the poles A and B, which can be still further intensified by the introduction of a Leyden jar. These sparks contain particles of the pole in the condition of glowing vapour. If the poles, therefore, consist of the metal to be examined, which may either be used in the form of a wire or in the form of irregular fragments fixed by means of clips, the sparks will exhibit the corresponding spectrum of the metal when seen through the spectroscope.

FIG. 112.

Induction apparatus.

This method of observation demonstrates that the representation of spectra upon the screen was inexact; each of the bright lines now shows itself to be composed of a number of extremely fine lines which, owing to the poor definition of the objective image, previously coalesced into a more or less broad band. Owing to the great number of fine bright lines, the spectra of the heavy metals are very complex. In the spectrum of iron, for example, more than 450 bright lines have been counted.

68. In the light of the electric spark, not only do

particles of metal detached from the poles glow, but also particles of the gas through which the spark passes. In the method of observation just described, therefore, the metallic spectrum is not pure, but is mingled with the spectrum of the atmosphere. This admixture cannot however occasion any error, providing the spectra which glowing gases themselves give are known.

In order to render a gas incandescent the discharge of an induction apparatus is allowed to pass through a so-called Geissler's tube (fig. 113), which contains the gas in question in a rarefied state. The two ends of the tube present dilatations into which platinum wires are fused. These wires are connected with the poles of an induction apparatus, and immediately a beautiful stream of light traverses the interior of the tube, the colour of the light varying with the nature of the contents. If the tube contain hydrogen the middle constricted portion shines with a splendid purple-red, the brilliancy of which is nevertheless too feeble to permit its spectrum to be projected upon a screen so as to be visible at any distance. If the tube be looked at through the spectroscope the light of the hydrogen appears to be composed of three homogeneous kinds of light: a red, a bluish green, and a violet line coming into view. (*See* Plate

FIG. 113.

Geissler's spectrum tube.

of Spectra, No. 12.) A tube filled with rarefied nitrogen shines with a peach-blossom colour, but gives a far more complex spectrum than that of hydrogen; for in the red, orange, yellow, and green, numerous closely approximated bright lines are seen separated from each other by slender dark lines; in the blue and violet, on the other hand, there are broad bright bands which are sharply defined towards the less refrangible side, but are gradually shaded off towards the refrangible side. (No. 13.)

Plücker and Hittorf, and more recently Wüllner, have demonstrated that in this method of observation different spectra are obtained with the same gas, if the presence of the gas and the kind of electrical discharge are appropriately altered. If with the induction apparatus a Leyden flask be connected, and the shock thus intensified be transmitted through the same tube containing nitrogen, light of another colour may be observed to be emitted from it, and if this be examined with the spectroscope it exhibits a spectrum consisting of many sharply-defined bright lines. A Geissler's tube filled with nitrogen thus gives two quite distinct spectra, according to the kind of electrical discharge. With low electric tension it gives the *spectrum of the first order*, consisting of bright striæ and bands, whilst with high tension it gives the *spectrum of the second order*, consisting of narrow bright lines. Other gases behave in a similar manner. Plücker and Wüllner have even shown that hydrogen, under increased pressure and with electric discharges of high tension, gives a continuous spectrum, and hence emits light of all degrees of refrangibility. In the same way Frankland has observed that a flame of hydrogen burning in oxygen under very high pressure emits white light, which gives a continuous spectrum. Our knowledge of the processes which take place in Geissler's tube during electrical discharges is still too imperfect to permit the conclusion to be drawn from the phenomena just described that the same gas can furnish different spectra. It is, on the contrary, not improbable that the spectra presenting lines (to which the above-mentioned hydrogen spectrum belongs) characterise the simple gases, whilst the spectra presenting bands belong to certain of their chemical compounds.

69. The spectra that have hitherto been considered may be arranged in the three following classes:—

1st, *Continuous spectra*, like those of the glowing carbon points in the electric lamp, Drummond's lime light, the magnesium light, white-hot platinum, iron in a state of fusion, and, speaking generally, and with but few exceptions, all white-hot solid or fluid bodies, whatever may be their composition. All these exhibit a spectrum which on beginning to be luminous presents the extreme red, and as the temperature rises constantly extends towards the more refrangible end, and finally becomes complete and continuous when white heat is attained. The flames of candles, lamps, and gas-burners also give continuous spectra, for they owe their brightness to the particles of solid carbon floating in them. Finally, to this group belong the above-mentioned continuous spectra which are observed under certain circumstances in gases.

2nd, Spectra which present a number of *bright lines and striæ on a dark background.* These are peculiar to *glowing vapours and gases,* each chemical element and chemical compound having its own characteristic spectrum. It is this which constitutes the basis of spectrum analysis.

3rd, The *solar spectrum* which exhibits a large number of fine *dark lines*—the lines of Fraunhofer—on *a bright ground*. These lines are perceived by means of the spectroscope in ordinary daylight, in the light of the moon, and in that of the planets, and hence not only in the direct but in the reflected light of the sun. The fixed stars, as independent suns, exhibit spectra which are similar but not identical with that of the sun. The circumstance that the dark lines of the fixed stars are not exactly coincident with those of the sun, permits the conclusion to be drawn that the lines of Fraunhofer, or at least a large number of them, do not proceed from any action of the atmosphere of our earth, but are peculiar to the solar light at its source. An endeavour must now be made to obtain more exact information in regard to the cause leading to their production.

CHAPTER XI.

SPECTRUM ANALYSIS OF THE SUN.

70. FRAUNHOFER first observed that the bright yellow line of the Sodium flame occupies the same position in the spectrum as the dark line, D, of the solar light. In order to demonstrate this, a right-angled prism (fig. 114) must be so placed in front of the slit which has hitherto been employed to throw the spectrum, that it only covers the lower half of the slit. From the side B the light of the electric arc, saturated with Sodium vapour, falls upon the prism, and undergoing total reflexion, is deflected by the oblique surface to the slit, whilst the sun's rays, as before, penetrate through the upper uncovered part. The spectra of the two sources of light corresponding to the two halves of the slit are therefore thrown upon the screen, one being immediately above the other, permitting them to be conveniently compared. It will then be seen that the *bright Sodium line forms the exact continuation of the dark line D in the solar spectrum*, and the conclusion may be drawn that the Sodium light possesses the same refrangibility as the

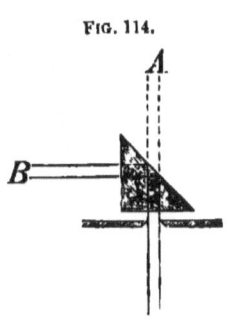

FIG. 114.

Action of the comparison prism.

line *D*. The 'Sodium light' and 'the *D* light' are therefore equivalent. (*See* Spectrum Plate, Nos. 1 and 5.)

Such a comparing—or comparison—prism may be applied to the slit of any spectroscope (fig. 115.) It permits the light coming from any source to be looked at coincidently with that of the solar spectrum, one occupying the upper, the other the lower half of the field of vision, and thus permits them to be directly compared. The solar spectrum, owing to the numerous fine lines it exhibits, may be taken as a scale by which all others may be measured.

By means of the comparing prism it may be demonstrated that the three bright lines of the hydrogen

FIG. 115.

Comparing prism at the slit of the spectroscope.

flame possess exactly the same refrangibility as three dark lines in the solar spectrum. The red line occupies precisely the position of the dark solar line *C*; the greenish blue corresponds to the line *F*, and the dark blue to a Fraunhofer's line which lies immediately in front of *G*. (*See* the Plate of Spectra, Nos. 1 and 12.)

Kirchhoff in like manner, in endeavouring to determine the precise position of the bright lines of metals, used the solar spectrum as a scale, and found that there were Fraunhofer's lines which corresponded to each of the iron lines he had observed. The coincidence descends to the minutest particulars; the more brilliant a bright iron line appears the blacker is the corresponding Fraunhofer's line; the more defined is the line of the metal the more definite is also the solar line; if, on the contrary, it be faint and have softened edges, there is a corresponding indistinctness in the solar spectrum. Thus every bright iron line (of which Ångström and Thalén have lately counted not less than 460), has its dark counterpart in the solar spectrum. The exact coincidence of so many bright iron lines with dark solar lines cannot be accidental. On the theory of probabilities millions of millions might be wagered to one that these lines have a common origin, or in other words, it is almost certain that both kinds of lines are produced by the glowing vapour of iron.

71. How does it happen, however, that the lines which in the spectrum of a glowing vapour appear *bright upon a dark ground* are seen conversely in the solar spectrum, *dark upon a bright ground*. A few experiments will show how an answer to this question may be given. The continuous spectrum of the electric light passing between the carbon points is projected upon the screen, and a fragment of Sodium is placed in the cavity of the lower pole. As it vaporises it invests the white-hot upper carbon point with a sheath of flame, which emits the well-known homogeneous yellow light. But there may now be seen upon the screen in the continuous spectrum *a dark line*, occupying exactly the position where before was the bright Sodium line, and where it now again immediately appears if the carbon poles be so far separated that the light of the arc of flame alone reaches the prism.

From this experiment the conclusion may be drawn

that the yellow sheath of flame permits all kinds of rays proceeding from the white-hot carbon to pass easily through it, with the exception of *that kind of ray which it emits itself*. This is completely arrested or absorbed; in other words, the vapour of Sodium is almost opaque for rays of its own kind, whilst it is perfectly permeable to all other kinds of rays.

Fig. 116.

Bunsen's apparatus for the absorption of Sodium light.

This peculiarity of the glowing vapour of Sodium may be very beautifully shown by means of an apparatus constructed by Bunsen (fig. 116). The flask A, closed by an elastic stopper perforated with three holes, contains a solution of common salt (sodium chloride), besides some sulphuric acid and zinc. From the mixture hydrogen gas is evolved, which carries with it small droplets of the solution of common salt. Coal-gas is conducted into the flask by means of the bent tube e, which, after admixture with the hydrogen gas containing solution of common salt, streams out through the tubes a and c. The coal-gas flame is almost non-luminous *per se*, but presents a yellowish tint from the admixture of the vapour of Sodium, and becomes mingled with air before undergoing combustion in the metal chimneys b and d. The chimney b widens like an inverted cone above, and from its semicircular slit-like aperture a broad extremely hot and bright Sodium flame is emitted. The other chimney, d, is funnel-shaped, and is provided above with a cover having an aperture in the centre.

SPECTRUM ANALYSIS OF THE SUN.

Incomplete combustion takes place in it, and a feeble flame, caused by the products, appears above the opening. This small Sodium flame appears almost perfectly dark upon the bright background of the large Sodium flame; and as it is almost opaque for Sodium light, it presents us with the surprising phenomenon of a *black flame* (fig. 117).

FIG. 117.

Absorption of the Sodium flame.

It cannot be doubted that the flame is not in itself black, but emits yellow Sodium light, as indeed may be immediately seen if the large flame is extinguished. As it appears dark upon the bright background, the quantity of light which it emits, together with that which it still transmits of the flame behind it, taken together, must be smaller than the intensity of the light of the posterior flame. It must thus, consequently, be less bright; or, since the intensity of light rises and falls with the temperature, less hot than the latter. Owing to the peculiar construction of these metal chimneys, the large flame is rendered as hot as possible, whilst the small one is reduced to as low a temperature as possible. If the anterior flame were bright enough to cover or even to surpass by its own luminosity the loss of light effected by absorption, the small flame would appear as bright or even still brighter than its background.

The dark Sodium line also which has heretofore been seen in the spectrum is not absolutely black; it still receives the sum of the D-light emitted and transmitted from the electric arc. It appears, however, in comparison with its environment—the brilliant spectrum of the carbon light-- dark.

The spectrum of Lithium can be similarly inverted to that of Sodium. For if a salt of Lithium be placed on the inferior charcoal point of the electric lamp, as well as a fragment of Sodium, Lithium and Sodium vapours must be coincidently present in the flame; and there is now seen in the spectrum, besides the dark line D, a dark line in the red exactly in the position where the bright red Lithium line was previously visible. The Lithium vapour thus absorbs just those rays which it itself emits.

The law which has been demonstrated in the case of Natrium and Lithium holds good generally. *Every gas and every vapour absorbs exactly those kinds of rays which it emits when in the glowing condition, whilst it permits all other kinds of rays to traverse it with undiminished intensity.*

This capability of absorbing remains unaltered under great variations of temperature, whilst the brilliancy of the light emitted rapidly increases or diminishes with the temperature. If therefore a source of light which gives a continuous spectrum be looked at with a spectroscope through a sheath of vapour, various appearances may be presented. If the vapour be so hot that it emits more light than it annihilates by absorption, its line-spectrum will be seen bright upon the less bright ground of the continuous spectrum. If its capacity of emitting light at a lower temperature be just sufficient to cover the loss of light caused by absorption, a continuous spectrum will be seen, and the presence of the vapour will scarcely be recognisable. Lastly, if at a still lower temperature the emitted light be insufficient to make up for that lost by absorption, the lines of the vapour will appear dark upon the bright ground of the continuous spectrum, or in other words, the inverse spectrum of the vapour or gaseous body is developed.

72. The inversion of gas spectra solves the enigma of Fraunhofer's lines, and at the same time gives an insight into the physics of the sun. The sun, as Kirchoff maintains, may be regarded as an extremely hot mass, whose glowing white-hot surface, the *photosphere*, emits white light, and in and by itself would give a continuous spectrum. Outside of the photosphere and surrounding the sun is an atmosphere of glowing gases and vapours, which is called the *chromosphere*; and this constituent, though of lower temperature than the photosphere, is still sufficiently hot to maintain heavy metals in the state of vapour. And since the light of the photosphere, before it reaches the earth, must traverse the chromosphere, it is subjected to the absorbing action of the gases and vapours found in it; and it is to this action that the lines of Fraunhofer owe their origin. The solar spectrum is consequently to be regarded as resulting from the juxtaposition of the *inverted spectra* of all those substances which are contained in the gaseous state in the solar atmosphere.

From the facts already mentioned it would appear that Hydrogen, Sodium, and Iron must be constituents of the solar atmosphere. Moreover, exact comparisons of the solar spectrum with the line-spectra of terrestrial substances show that a series of other elements [*] exist in the sun. Thus, for example, the two lines *H* are produced by Calcium vapour, and the group indicated

---

[*] The presence of the following elements has been demonstrated with certainty in the solar atmosphere:—Sodium, Calcium, Barium, Magnesium, Iron, Chromium, Nickel, Copper, Zinc, Strontium, Cadmium, Cobalt, Hydrogen, Manganese, Aluminium, and Titanium.

by Fraunhofer with *b* are produced by the vapour of Magnesium. The line *G* depends upon Iron, and partly also the group *E*. The lines *C* and *F* belong, as we already know, to Hydrogen, and *D* to Sodium. But besides these there are a number of dark lines in the solar spectrum which do not correspond to any known terrestrial element. In addition to the lines of Fraunhofer, indubitably belonging to the sun, there are many other dark lines in the solar spectrum which originate from the absorptive action of the terrestrial atmosphere, and are therefore called *atmospheric lines*. That they are really produced by the atmosphere is easily recognised by the fact that they are seen more distinctly or even first make their appearance when the sun approaches the horizon, and when consequently its rays have to traverse a much greater extent of the terrestrial atmosphere. The Fraunhofer's lines *A* and *B*, the darkness of which essentially depends on the relative position of the sun, must on this account be regarded as atmospheric.

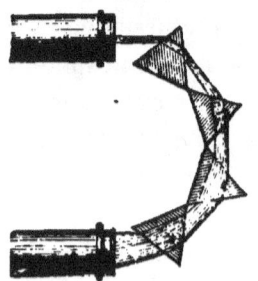

Fig. 118.

Telescopes with four prisms.

To make a comparison of the spectra of metals with that of the sun with the precision required for this kind of investigation, a spectroscope with only one prism is, on account of its small dispersive power, insufficient. Kirchhoff, therefore, in order to obtain an exact drawing of the solar spectrum to compare with the lines of metallic elements employed a spectroscope with four prisms, appropriately arranged behind one another (fig. 118), together with a highly magnifying telescope. By this instrument fresh lines are rendered visible, and groups of lines, which in weaker instruments appear only as misty striæ, are resolved into their several lines. Thus, for example, the dark line D can be shown by its means to be composed of two fine lines, D_1 and D_2, as was already known to Fraunhofer; and in the same way the bright line of Sodium splits into two lines which correspond in the most precise manner with two solar lines. The excellent drawings of the solar spectrum made by Kirchhoff and Hoffman, and subsequently by Ångström and Thalén, are as important and indispensable for the spectrum analysis of the sun and celestial bodies as the chart of stars is to the astronomer for enquiry into the position of the fixed stars.

73. If the explanation of the lines of Fraunhofer given by Kirchhoff be correct, those parts of the solar atmosphere which project at the edge of the sun beyond the photosphere should exhibit bright lines in the spectroscope in place of the dark Fraunhofer's lines.

The so-called protuberances afforded an instant and crucial test of the truth of Kirchhoff's hypothesis. In total eclipses of the sun, at various points of the sun's edge reddish projections appear, which sometimes resemble clouds, sometimes hook-like curved horns, and sometimes snowy mountains glowing with the rosy tint of evening. In the uneclipsed sun these protuberances cannot be seen, because their feeble light is lost in the brightness of the terrestrial atmosphere during the day. The first spectrum of the protuberances was obtained during the solar eclipse of August 18, 1868. *It presented bright lines,* amongst which the three lines of Hydrogen (C, F, and one a little in front of G), and a yellow line behind the double line D, which corresponds neither to a Fraunhofer's line, nor to that of any known terrestrial substance, and which has been since designated D_3, are the most conspicuous. It was thus demonstrated that the protuberances are gaseous, and that they are principally composed of hydrogen.

Janssen, who was sent to the East Indies by the French Academy of Sciences to observe this eclipse, discovered on the following day a method of seeing the bright lines of the protuberances without any eclipse of the sun, and when shining at its brightest. The idea of this method had previously been suggested by Lockyer, and had been carried into effect by him before he had received information of the discovery made by the French observer.

The reason that we are unable to see the protuberances with an ordinary telescope in bright sunshine is on account of the great brightness of the terrestrial atmosphere, rendered luminous by the sun, which overpowers the feeble light of the protuberances. In order that the spectrum of the protuberances should be seen, it is necessary to lower the light of the terrestrial atmosphere to a sufficient degree, yet without at the same time materially weakening that of the protuberances.

The practicability of effecting this depends on the great difference that exists between ordinary daylight and the light of the protuberances. The former consists of all possible kinds of rays, and gives, apart from Fraunhofer's lines, a continuous spectrum; the latter, on the other hand, consists of only a few homogeneous kinds of light, to which, in its spectrum, the previously-mentioned bright lines correspond. By multiplying the prisms of the spectroscope the continuous spectrum of ordinary daylight may be indefinitely extended, and its brilliancy so far diminished that it is scarcely to be perceived. By the same system of prisms the bright lines of the spectrum of the protuberances may indeed be separated widely from one another, but are not materially weakened in brilliancy. In order, therefore, to see them distinctly upon the dark ground of the almost imperceptible spectrum of the atmospheric light, it is only requisite to use a *strongly dispersing spectroscope.*

Were the spectroscope pointed directly towards the sun, light from all its parts would simultaneously penetrate the slit of the instrument and the ordinary solar spectrum would be produced; but with the present object in view it is necessary that each segment of the sun should be investigated separately. This object is attained by placing a spectroscope instead of the ocular in a telescope, and receiving the small image of the sun formed at the focus upon the plane of the aperture of the slit. By this means any given part of the sun's disk or edge can be made to fall separately upon the slit.

This arrangement renders it possible not only to recognise by its bright lines the presence of a protuberance, but also to see its complete form with well defined borders. If we make, for example, the slit so wide that it takes in the whole image of a protuberance between its borders, we see through the spectroscope as many images of it as there are homogeneous rays in the light of the protuberance. These images are quite sharply defined, and in consequence of the great dispersion of the spectroscope, are so widely separated from each other that only one is seen in the field of vision, and the protuberance can be seen at will, red by virtue of its C rays, or greenish blue by its F rays. This method of observation cannot be applied to a white object, because the innumerable coloured images would be arranged and become confused in a continuous series. The protuberances are to be regarded as violent eruptions of gases, which are shot forth to an extraordinary height above the proper solar atmosphere (chromosphere) and it is to their absorptive power that the Fraunhofer's lines are due. In the eclipse of December 22, 1870, the American observer Young also perceived the bright lines of the chromosphere itself. He made the following report upon this important observation, which powerfully supports Kirchoff's view: 'As the solar sickle became narrower, I remarked how all the dark lines became progressively fainter, but I was wholly unprepared for the extraordinary phenomenon which in an instant presented itself to my eye at the moment when the dark disk of the moon entirely covered the photosphere of the sun. The whole field of vision was filled with bright lines which suddenly appeared with the greatest brilliancy and then again vanished, so that after the lapse of scarcely two seconds nothing remained of those lines which had just been the object of my investigation. It is obviously impossible for me to state with certainty that all the bright lines which filled the field of vision occupied exactly the same position as the lines of Fraunhofer, but I am convinced that it was so, for I recognised various groups of lines, and the whole disposition, as well as the relative intensity of the spectrum, seemed quite familiar to me.'

Since this observation, which was made during an eclipse, the bright lines of the chromosphere have been seen in bright sunshine by means of the same method of research as that above detailed for examining the protuberances. Young has in this way observed not less than 273 bright lines in the chromosphere, of which 64 belong to Iron.

Spectrum analysis has been applied with the greatest success, not to the sun alone but to other celestial objects. It is impossible, however, to go into farther detail in regard to the results obtained, since this subject is beyond the limits assigned to this work.

CHAPTER XII.

ABSORPTION.

74. THAT gaseous bodies are capable of producing absorption lines not only in the incandescent condition, but at far lower temperatures, is shown by the above-mentioned atmospheric lines of the solar spectrum, which are essentially due to the aqueous vapour contained in the air. Other gases possess a similar power of absorption, two examples of which may here be mentioned.

After the spectrum of the electric light has been thrown upon the screen, a small test-tube, containing some nitric acid and copper, is placed in front of the slit. As the acid dissolves the metal, a yellowish-red gas is developed, through which the rays of light must pass before they reach the prism.

It may now be seen (fig. 119, 1) that the previously continuous spectrum is interrupted by innumerable dark lines (Brewster has counted about 2,000), which closely resemble the lines of Fraunhofer. They are sparingly present in the red part, but are more closely arranged towards the violet end, and render it quite faint.

If a little Iodine be volatilised in another test-tube, and the light of the electric lamp be transmitted through the beautiful violet vapour, the spectrum may

ABSORPTION. 173

again be observed to present a number of dark lines
(fig. 119, 2), which, however, have a very different
arrangement from the above. They are principally
situated in the orange, yellow, and green; and indeed

FIG. 119.

Absorption spectra of nitrous oxide and of the vapour of iodine.

are so closely grouped in the latter that they quite
darken it. On the other hand, the blue and violet part
of the spectrum is quite free from them. This absorption spectrum, as Wüllner has shown, is exactly the
converse of the spectrum of glowing Iodine vapour.
If, for example, the reddish-yellow light of a hydrogen
flame, saturated with Iodine vapour, be examined
through the spectrum apparatus, bright lines are obtained at those points where the absorption spectrum
appears dark.

The reddish-yellow colour of the nitrous acid, and
the violet colour of the vapour of iodine, are the necessary consequences of their peculiar powers of absorption;
for as the nitrous acid arrests certain kinds of rays of
the white light traversing it, and especially the violet
ones, the mixture of the rest is no longer white, but
just the reddish-yellow tone of colour proper to this
gas. For the same reason Iodine vapour, being almost
opaque for the yellow and green rays, exhibits a mixed

tint, formed by the red, blue, and violet rays which it transmits, and which appear violet to our eyes.

75. The different colours of transparent solid and fluid bodies similarly result from their peculiar capabilities of absorption, a series of examples of which may now be given. When a solution of *permanganate of potash* contained in a glass trough with parallel walls is placed in front of the slit of the Heliostat,* (fig. 120), the red and blue-violet regions of the spectrum appear unaltered, whilst the yellow and the green appear darkened, and upon the dark ground are fine black striæ. It is unnecessary that any explanation should here be entered into of the mode in which the reddish-violet colour of the fluid results from this phenomenon of absorption.

If again *blood* diluted with water be placed in the glass trough, the violet end of the spectrum vanishes, and between D and E two broad dark bands (fig. 120, 2) make their appearance. The red colour of blood is thus not a simple colour, but a mixture of all those colours which still remain over in its spectrum. The slightest chemical alteration in blood betrays itself immediately by a corresponding change in the spectrum. Thus poisoning by carbonic oxide gas (fire-damp), or by hydrocyanic acid, may be immediately recognised by the changed appearance of the blood spectrum. The spectroscope may thus render important services to Physiology and Forensic Medicine.

Plants owe their green colour to the '*chlorophyll*'

---

* If these experiments are made with the light of the sun, the Fraunhofer's lines are seen in addition to the absorption phenomenon and furnish satisfactory points of comparison for the determination of the position of the absorption lines

ABSORPTION. 175

contained in their cells. An alkaline solution of this colouring material gives a highly characteristic spectrum (fig. 120, 3). In the middle of the red is a deep black band, which occupies the interspace between the lines

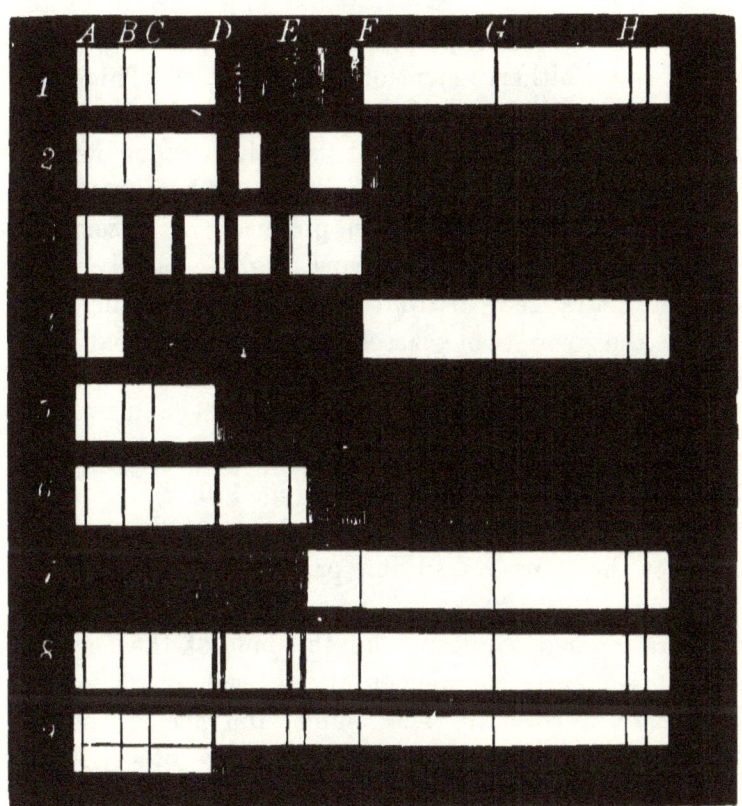

Fig. 120.

Absorption spectra.

B and C; three feeble absorption striæ are seen in the orange-yellow and green; the indigo-violet part of the spectrum from F onwards is completely absent.

If a piece of *glass coloured blue with Cobalt* be held in front of the prism, the spectrum shown in fig. 120, 4, is obtained. In this the whole tract from *B* to *F* is shaded, with the exception of a feebly luminous line in the yellow-green. The extreme red, on the other hand, before *B*, as well as the entire indigo-violet extremity of the spectrum, remains unchanged.

A glass coloured red with oxide of Copper gives an absorption spectrum of a far more simple kind than any of those hitherto mentioned (fig. 120, 5). This kind of glass only allows the red and orange-red rays as far as *D* to pass through it; it is quite opaque for all other colours. If a red glass be placed before a blue cobalt glass the combination produces by absorptive action a nearly homogeneous light, namely, the extreme dark red in front of *B*, which is the only colour that the two glasses are together capable of transmitting.

A solution of *Potassium bichromate* is only transparent for the less refrangible part of the spectrum as far as to the Fraunhofer's line *b* (fig. 120, 6). A solution of the *ammoniated oxide of Copper* is transparent only for the more refrangible part, from about the line *b* onwards (fig. 120, 7). The orange-yellow colour of the first-named solution, and the blue of the second, are consequently complementary to each other. Two glass cells filled with these fluids, and placed one behind the other, scarcely permit the passage of any light. The one fluid looked at through the other appears completely black. Nevertheless absorption does not always produce the particular tone of the transmitted light. If only a very small extent of the spectrum be absorbed, the mixture of the transmitted rays does not differ remarkably from white. As an example of this, a piece of glass may be adduced which contains in a state of chemical combination the rare metal Didymium. To the naked eye it appears nearly colourless, but if it be brought in front of the slit, two thin black striæ appear in the spectrum at the line D, and two less well-marked ones in the green at E and b (fig. 120, 8), which are so characteristic of Didymium that they enable the smallest quantity of this metal in solution to be detected. If the solid oxide of Didymium be heated to incandescence, *bright lines* appear in the spectrum of the emitted light in place of the dark lines. We have thus in Didymium an example of a solid which when incandescent does not give a continuous but a linear spectrum. The oxides of the metals Erbium and Terbium, which are also rare, behave in a similar manner.

If an absorbing substance be employed in a progressively thicker layer or in a greater degree of concentration, the absorption bands become, without changing their position, broader and darker, and colours which were previously transmitted gradually disappear. Thus it comes to pass that with increasing thickness or concentration the tone of colour of the transmitted light frequently becomes quite different. To demonstrate this a number of gelatine disks coloured with litmus may be used, which are placed between two colourless glass plates in a graduated manner. If these be placed before the slit, there will be seen in the spectrum (fig. 121) the graduated amount of absorption corresponding to the different thicknesses of the gelatine. In the case of the thinnest layer only a thin dark band is seen in front of D, whilst the thickest layer only permits the red end of the spectrum to be seen. The appearance of this spectrum explains why a layer of litmus gradually increasing in thickness first appears whitish, then blue, then violet, and finally purple-red. Similarly a solution of chlorophyll, which in a thin layer appears green, transmits when very thick only the extreme dark-red rays.

Fig. 121.

Absorption of the colouring matter of litmus with different thicknesses of the layer.

The absorption spectra being thus not less characteristic in demonstrating the presence of the bodies to which they belong than are the spectra of the light emitted from glowing vapours, spectral analysis opens up a wide field of application. The discovery of adulteration of colouring matters and of food may be particularly mentioned in practical life.

76. In the experiments hitherto made the rays emerging from the prism have been received upon a paper screen because the rough surface of the paper reflects* the different coloured rays diffusely, enabling the complete spectrum to be seen on all sides. Instead of the usually perfectly white screen, another one may be selected, the upper half of which is covered with white and the lower half with red paper. The screen must be placed in such a position that the horizontal line of junction of the two papers halves the spectrum throughout its whole length. In its upper half, which

* See §§ 8 and 15.

falls upon the white paper, the spectrum exhibits all the colours as clearly as before, but in the lower half, which falls on the red paper, the colours yellow, green, blue, and violet are almost entirely absent, whilst the red and orange are almost as bright as when they fall on the white screen (fig. 120, 9).

This experiment proves that the red paper possesses in a high degree the power of reflecting diffusely the red and orange-coloured rays, but that it does not reflect the other kinds of rays falling upon it, but, on the contrary, swallows them up, or, as we say, absorbs them. It is obvious therefore why this paper appears red when illuminated by the white light of day.

If this experiment be repeated with yellow, green, and blue paper successively, it will be found that each absorbs other parts of the spectrum, and that the particular colour which it possesses in daylight is always the tint caused by mixture of all those rays which it diffusely reflects.

White paper absorbs no one of the homogeneous colours present in the light of the sun in particular, but reflects all in their original state of mixture, and it is on this account that it appears by daylight white. A surface is called *grey* which possesses an equally small power of diffusion for all colours. Lastly, everything appears *black* the surface of which is of such a nature that all kinds of rays are absorbed by it.

The whole range of colours presented by objects in all their variety may thus be explained on the principle of absorption. All objects, whether seen by transmitted or by reflected light, exhibit exactly that colour which is complementary to the sum of the rays absorbed.

The bright fresh green of plants, for example, results from the absorbing action of chlorophyll, and has therefore the same composition as the light passing through a solution of chlorophyll (see fig. 120, 3). It contains, namely, the extreme red in front of the Fraunhofer's line B quite undiminished in intensity, the orange-yellow and green between C and E with tolerably strong brilliancy, and a little blue, but the middle part of the red (corresponding to the absorption striæ between B and C) as well as the indigo and violet from the middle between F and G, are almost completely absent.

This peculiar composition of the green colour of plants explains the surprising appearance which a well-wooded landscape presents on a sunny day if looked at through two properly selected glass plates, of which one is a blue cobalt glass whilst the other is faintly tinted with oxide of copper. Spectacles made of these two glasses superimposed on one another (*erythrophytoscope*) permit only the extreme red constituent of the green colour of plants, with some blue-green and blue but no green or yellow, to reach the eye. The foliage of plants is therefore seen coloured of a beautiful red, whilst the bright sky is of a deep violet-blue colour, the clouds of a delicate purple, and the earth and rocks of a violet-grey.

77. The essential nature of the colours of objects may thus be strikingly indicated, by saying that they are the residue of the light by which they are illuminated after abstraction of those rays which are extinguished by absorption. It follows as a matter of course that *objects can only exhibit such colours in transmitted as well as in diffusely reflected light as are already contained in the incident light.* Hence in order that a

sheet of red paper should appear red, red rays must be contained in the light by which it is illuminated. The light of day contains such rays. But if the room be darkened and the paper illuminated with the monochromatic yellow flame of Sodium, it appears black.

With homogeneous illumination differences of colour are no longer perceptible. The variations of light and shade are alone visible. Hence the wreath of flowers which appeared so luxuriant in the above experiment would, when seen with homogeneous light, seem withered and yellow; and a picture, rich as it might really be in colour, would resemble a sepia drawing. Were the sun a sphere of glowing vapour of Sodium, all terrestrial nature would present this monotonous and gloomy aspect. It requires the white light of the sun, in which innumerable colours are blended, to disclose to our eyes the variegated tints of natural objects. And so again, if a Magnesium wire be held in the Sodium flame, its white light, as by a stroke of magic, restores the fresh colours to the wreath of flowers, to the picture, and everything around.

The light of gas and candles contains all the colours of the solar spectrum, though not mixed in exactly the same proportion. The yellow rays are very abundant, whilst the blue and violet are relatively much less abundant than in solar light. This affords an explanation of the well-known fact that green and blue clothing materials are difficult to distinguish by candlelight. For green materials reflect especially the green and a few blue rays; blue materials, in addition to the green, the blue rays especially; but since blue is only sparingly present in candlelight, whilst green is abundant, objects presenting both colours by daylight appear more or less of a green colour by candlelight.

If the two colours are mingled the mixture presents that colour which remains over after the abstraction of all the rays absorbed by the two materials. It is, for example, generally known that a mixture of blue and yellow, as of Prussian blue and gamboge, produces a green. This is by no means in opposition to the fact above stated (§ 57), that the yellow and the blue of the spectrum unite to form white. For in order that our eyes should receive the impression of white it is necessary that blue and yellow rays should enter them simultaneously. A mixture of Prussian blue and gamboge emits neither blue nor yellow, but essentially green rays. The former colouring matter absorbs the red and yellow, the latter the blue and violet rays, and the green rays therefore alone remain in the diffuse light reflected from the mixture.

CHAPTER XIII.

FLUORESCENCE. PHOSPHORESCENCE. CHEMICAL ACTION.

78. THE question now arises, what becomes of the rays that have undergone absorption? Are they in fact, as they appear to be, annihilated? A series of phenomena now to be considered will give us an answer to these questions.

If water containing a little *Æsculin*, a substance contained in the bark of the horse chestnut in solution, be placed in a flask, and the rays of the sun or of the electric lamp concentrated by a lens situated at about its focal distance from the vessel, be directed upon it, the cone of light thrown by the lens into the interior of the fluid will be seen to shine with

Fig. 122.

Fluorescence.

a lovely sky-blue tint. The particles of the solution of Æsculin in the path of the beam become spontaneously luminous, and emit a soft blue light in all directions. The cone of light appears brightest at the point where it enters into the fluid through the glass, and quickly diminishes in brilliancy as it penetrates more deeply.

There are great numbers of fluid and solid bodies

which become similarly self-luminous under the influence of light. This peculiarity was first observed in a kind of spar occurring at Alston Moor in England, which, itself of a clear green colour, appears by transmitted solar light of a very beautiful indigo-violet colour. From its occurrence in Calcium fluoride the phenomenon has been named *fluorescence*.

In order to understand more precisely the circumstances under which fluorescence occurs, the solution of Æsculin must again be referred to. The light *before* it reaches the lens must be allowed to pass through just such another solution of Æsculin contained in a glass cell with parallel walls. The cone of light proceeding from the lens, as long as it passes through the air, does not appear to have undergone any material change, it is just as bright and just as white as before. In the interior of the fluid however it *no longer presents a blue shimmer but becomes scarcely perceptible.*

Thus it is seen that light which has traversed a solution of Æsculin is no longer capable of exciting fluorescence in another solution of Æsculin. Those rays consequently which possess this property must be arrested by the first solution of Æsculin. Similar results are obtained in the case of every other fluorescent substance.

The general proposition can therefore be laid down, *that a body capable of exhibiting fluorescence fluoresces by virtue of those rays which it absorbs.*

In order to determine what rays in particular cause the fluorescence of Æsculin, the spectrum must be projected in the usual way; but instead of its being received upon a paper screen it must be allowed to fall upon the wall of a glass cell containing a solution of

FLUORESCENCE. 185

Æsculin, that is to say, upon the solution itself, and it must then be observed in what parts of the spectrum the blue shimmer appears. The red and all the other colours consecutively down to indigo appear to be absolutely without effect. The bluish shimmer first commences in the neighbourhood of the line G, and covers not only the violet part of the spectrum, but *stretches far beyond the group of lines H* to a distance which is about equal to the length of the spectrum visible under ordinary circumstances.

FIG. 123.—Solar spectrum with the ultra-violet portion.

From this the conclusion must be drawn that there are rays which are still more refrangible than the violet, but which in the ordinary mode of projecting the spectrum are invisible; these are termed the *ultra-violet rays*. They become apparent in the Æsculin solution because they are capable of exciting the bluish fluorescent shimmer in it. If sunlight have been used in the above experiments the well-known Fraunhofer's lines appear upon the bluish ground of the fluorescing spectrum, not only from G to H, but the ultra-violet part also appears filled with numerous lines, the most conspicuous of which are indicated by the several letters L to S (fig. 123). That these lines, like the ordinary Fraunhofer's lines, belong properly to solar light, and do not depend upon any action of the fluorescing substance, is evident from the circumstance that with the electric light they are no more apparent in the ultra-violet than in the other colours, and further, because the same lines are seen in the solar spectrum, whatever may be the fluorescing substance under examination.

Quartz has the power of transmitting the ultra-violet rays far more completely than glass. If therefore the glass lens and prism hitherto used for projecting the spectrum be replaced by a quartz lens and prism, the ultra-violet part of the spectrum is rendered much brighter and is extended still further than before.

The ultra-violet rays of the spectrum can, moreover, be seen without the intervention of any fluorescing substance through a glass, or still better, through a quartz prism, if the bright part of the spectrum between B and H be carefully shut off. With feeble illumination its colour appears indigo-blue, but with light of greater intensity it is of a bluish-grey tint (lavender). The ultra-violet rays thus ordinarily escape observation, because they produce a much feebler impression on the human eye than the less refrangible rays between B and H.

An explanation is thus afforded why the solution of Æsculin, apart from its absorption, is colourless when seen by transmitted light; for since it absorbs only the feebly luminous violet and the entirely imperceptible ultra-violet rays, the mixed light that has passed through it still appears white and is not rendered materially fainter.

79. If the solar spectrum be thrown in the above-mentioned manner upon the fluid, its fluorescing part everywhere exhibits the same bluish shimmer; and spec-

troscopic examination shows that this bluish light has always the same composition, whether it is excited by the G rays or by the H rays or by the ultra-violet rays, and that it is formed of a mixture of red, orange, yellow, green, and blue. It is thus seen that the different kinds of homogeneous light, as far as they are generally effective, produce *compound* fluorescent light of identical composition, the constituents of which nevertheless are collectively *less refrangible than, or are at most equally refrangible with,* the exciting rays.

Amongst other fluorescing bodies may be mentioned the solution of Quinine, which is as clear as water, and has a bright blue fluorescence; the slightly yellow Petroleum, with blue fluorescence; the yellow solution of Turmeric, with green; and the bright yellow glass containing Uranium, which fluoresces with beautiful bright green fluorescence. It admits of easy demonstration that in these bodies also it is the more refrangible rays that call forth fluorescence. For if we illuminate them with light which has passed through a red glass no trace of fluorescence is visible. But if the red be exchanged for a blue glass the fluorescence becomes as strongly marked as with the direct solar light. A remarkable phenomenon is presented in the splendid bright green light which is emitted by Uranium glass under the action of blue illumination.

The highly refrangible rays which possess in so high a degree the power of exciting fluorescence are contained in large proportion in the light emitted by a Geissler's tube (*see* § 68) filled with rarefied nitrogen. In order to expose fluorescing fluids to the influence of this light the arrangement represented in fig. 124 may be employed with advantage. A narrow tube is surrounded by a wider glass tube, into which the fluid is introduced by a side opening which can be closed if required. Another form of Geissler's tube is represented in fig. 125, which contains in its interior a

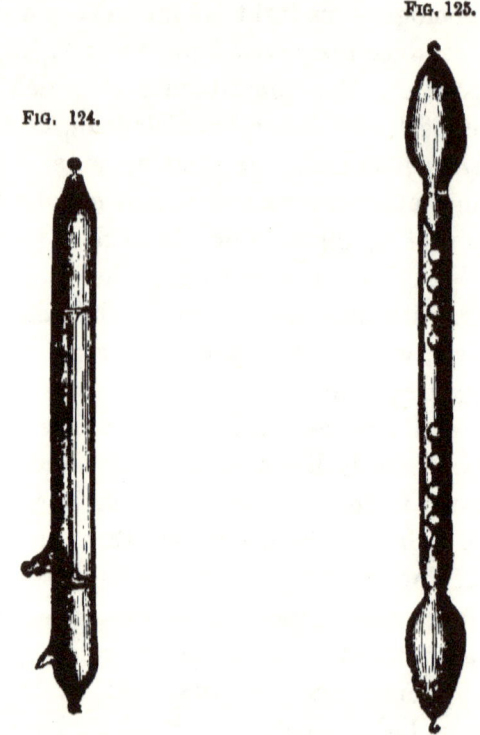

FIG. 124. FIG. 125.

Geissler's fluorescence tube.   Geissler's tube with Uranium glass spheres.

number of hollow spheres composed of Uranium glass. Where a beam of the reddish violet nitrogen light traverses the tube the Uranium glass balls shine with a beautiful bright green fluorescent light.

The electric light passing between carbon points is rich in rays of high refrangibility, indeed the ultraviolet end of its spectrum reaches even further than that of the solar spectrum. In the light of the Magnesium lamp the ultra-violet rays are also abundant, and both sources of light are therefore particularly well adapted to produce fluorescence, whilst gas and candlelight are nearly inoperative on account of the small amount of the more refrangible rays they contain.

80. It would nevertheless be incorrect to infer from the above facts that the more refrangible rays are exclusively capable of exciting fluorescence. A red fluid which is an alcoholic solution of Naphthalin red (Rose de Magdala, an anilin colouring material) and which even in ordinary daylight fluoresces with orange yellow tints of unusual brilliancy, will serve to demonstrate that even the less refrangible rays are capable of producing this effect. In fact, if the spectrum be projected upon the glass cell containing the fluid (fig. 126, 2), the yellow fluorescent light will be seen to commence at a point intermediate to C and D, and therefore still in the red, and to extend over the whole remaining spectrum as far as to the ultra-violet. The strongest fluorescence by far is shown behind the line D in the greenish-yellow rays. It then again diminishes, and becomes a second time more marked between E and b, from thence onward the fluorescence becomes fainter, then increases again in the violet, and gradually vanishes in the ultra-violet. In Naphthalin red, therefore, there are rays of low refrangibility, namely, the green-yellow rays behind D, by which its fluorescence is most powerfully excited.

The fluorescing spectrum received upon the fluid shows, as we have already mentioned, three regions of stronger fluorescence, and the absorption spectrum of Naphthalin, which by placing a small cell filled with the solution in front of the slit may be obtained upon a paper screen, gives a key to the cause of this phenomenon. In this spectrum (fig. 126, 1) a completely black band is visible in the green-yellow behind *D*, a dark band

FIG. 126.

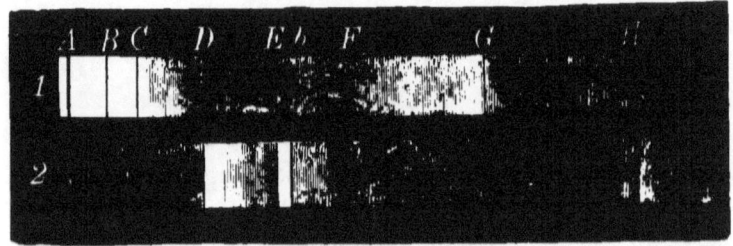

Absorption and fluorescing spectrum of Naphthalin red.

between *E* and *b*, whilst the violet end appears shaded. On employing a very strong solution of the Naphthalin colouring material, the whole spectrum vanishes with the exception of the red end, which remains apparent to a point behind *C*. If now the absorption spectrum be compared with that thrown upon the fluid, the intimate relation between absorption and fluorescence that has already been pointed out in the Æsculin solution is corroborated in the minutest particulars. *For every dark band in the absorption spectrum corresponds to a bright band in the fluorescing spectrum.* Every ray absorbed by the fluid occasions fluorescence, and the fluorescent light produced by it is the brighter the more completely the ray is absorbed.

A second example of the excitation of fluorescence by rays of small refrangibility is exhibited by a solution of chlorophyll. The spectrum projected upon this green fluid fluoresces of a dark red colour, from *B* to a point within the ultra-violet, exhibiting at the same time bright bands which correspond with the dark bands in the absorption spectrum (fig. 120, 3). Between B and C, where the greatest amount of absorption occurs, the fluorescence is also the most marked. But it is the middle red rays which here act most powerfully as excitants. It is remarkable that the red fluorescent light which the chlorophyll solution emits likewise lies, in regard to its refrangibility, between B and C. Chlorophyll solution affords a proof that all rays of the spectrum, with the exception of the extreme red in front of B, are capable of calling forth fluorescence. Their capacity for doing so depends simply on the power of absorption of the fluorescing substance. The most refrangible violet and ultra-violet rays are, however, characterised by the circumstance that they are capable of exciting all known fluorescing bodies.

81. Fluorescent light is only perceived so long as the fluorescent substance is illuminated by the exciting rays. As soon as the light falling on it is obstructed the coloured shimmer vanishes. It is only in the case of some fluorescing solid substances, as for example, Fluor-spar and Uranium glass, that, with the aid of appropriate apparatus (Becquerel's Phosphoriscope), a very short continuance of the fluorescence may be observed to take place in the dark.

There are, however, a number of bodies which, after being excited to self-luminosity by a brilliant light, continue to shine for a certain time in the dark. A series of pulverulent white substances, namely, the sulphur compounds of Calcium, Strontium, and Barium (which should be kept in hermetically sealed glass tubes), do not exhibit the faintest light in a dark room. Moreover, if they be covered with a yellow glass and illuminated with the light of a Magnesium lamp, they remain as dark as before. But if the yellow be exchanged for a blue glass, and the Magnesium light be allowed to play upon them for a few seconds only, they emit in the dark a soft light, each powder having its own proper tint of colour. This power of shining in the dark *after having been exposed to light is termed phosphorescence.* The property is possessed in a high degree not only by the above-named artificially prepared substances, but by various minerals, as the diamond, fluor-spar, and a variety of fluor-spar called Chlorophane.

Phosphorescence, like fluorescence, is an effect of *absorbed light*. For the refrangible rays which, in accordance with the results of the experiments that have been made, are alone capable of exciting these substances to self-luminosity are exactly those which they absorb. Phosphorescent light itself, examined spectroscopically, is found to be composed of rays the refrangibility of which is smaller than that of the exciting rays, and it is indeed compound even when the exciting light is homogeneous. The affinity between phosphorescence and fluorescence which expresses itself in this relation is unmistakable. Phosphorescence may be described as fluorescence which is prolonged for a certain length of time beyond the action of the exciting rays.

A remarkable fact discovered by Becquerel must not here be passed over in silence. When a card covered with Strontium sulphide is made feebly phosphorescent by daylight, and a solar spectrum is then projected upon it in a dark chamber, we observe in the course of a few seconds after the opening in the shutter has been closed that the whole surface of the card still continues to shine, with the exception of that part on which the less refrangible portion of the spectrum from A to F previously fell. In that part no phosphorescence is visible. The less refrangible rays are thus not only incapable of exciting phosphorescence, but they appear *even to destroy or disturb the phosphorescence called forth by the more refrangible rays.*

In order to avoid misunderstanding, it must further be remarked that the light of phosphorus (apart from the similarity of the name), the light of touchwood, of fire-flies, of various marine animals, etc., does not belong to the class of phosphorescent phenomena caused by the absorption of light which we have here considered. These bodies are rather to be regarded as *self-luminous* in consequence of chemical and physiological processes.

82. The nature of the substances exhibiting fluorescence or phosphorescence owing to the rays of light they have absorbed is in no way altered. There are, however, a number of bodies which undergo a permanent change in their nature—an alteration of their chemical composition—from exposure to light. Everyone must be familiar with numerous examples of this chemical action of light from the phenomena of daily life, and it is only necessary to mention such cases as the bleaching of linen and of wax, the fading of coloured stuffs, and the blanching of water-colour drawings.

How powerfully the chemical action of light can be exerted under certain circumstances may be shown by the following experiment. A mixture of equal parts of Chlorine and Hydrogen is introduced into a thin glass ball. If this be exposed to the daylight the two gases gradually combine to form Hydrochloric acid gas, a chemical compound the aqueous solution of which is generally known under the name of Muriatic or Hydrochloric acid. But if the light of the Magnesium lamp be allowed to fall on the sphere it instantly bursts with a loud explosion, and is broken into a thousand fragments; that is to say, under the influence of this brilliant light the chemical combination of the two gases and the associated development of heat takes place with such suddenness that the thin glass is unable to resist the pressure exerted.

If a yellow glass be placed in front of the Magnesium lamp, and the yellow light transmitted, which contains only the less refrangible rays of the spectrum, be allowed to act upon another of these little glass balls filled with the same mixture of gases, the ball will not explode, but it bursts directly if the yellow be exchanged for a blue glass. The conclusion therefore may be drawn that it is only the more refrangible rays of the spectrum that are capable of inducing the chemical combination of Hydrogen with Chlorine.

Whilst in the example just given the rays of light induce the *chemical combination* of two elementary substances, in other cases they can effect the decomposition of compound bodies. This is pre-eminently the case with the salts of silver on which Photography depends. The photographic process consists in receiving the image thrown by a camera obscura upon a glass plate covered with a layer of a sensitive preparation of silver, and as the silver salt is only decomposed when it is exposed to the light, and in proportion also to the brilliancy of the light, a permanent image is fixed upon the plate.

PHOSPHORESCENCE.

Daily experience shows that the more refrangible rays are more active in producing photographic effects than the less refrangible; a blue coat, for example, comes out very bright in a photograph, a red, on the other hand, very dark; although, looked at directly, the former appears to the eye the darker of the two. The most immediate key to the action of the different kinds of rays is obtained when we photograph the solar spectrum itself. The red, yellow, and the greater part of the green rays are then seen to be completely without action, whilst the blue, violet, and especially the ultra-violet part of the spectrum are depicted sharply with all their dark lines. Photography acts upon the ultra-violet rays still more than fluorescence; it constitutes a means not only of making this part of the spectrum visible, but also of fixing it permanently.

These groups of more refrangible rays, namely, the blue, violet, and ultra-violet, may fairly be characterised by the term 'photographic rays.' When, as is frequently done, they are called 'chemical rays,' the exclusive power is incorrectly ascribed to them of acting chemically. Their chemical action does not depend, as might be inferred from the term 'chemical rays,' upon any special chemical, or as it has also been called 'actinic' property inherent in them in opposition to the other rays, but simply upon the circumstance that all easily decomposed salts possess the property of absorbing the more refrangible rays whilst they allow the less refrangible to pass through them.

That the less refrangible rays are really capable of exerting a chemical action was demonstrated by H. Vogel. By the addition of certain anilin colouring matters to bromide of silver he was able to produce photographic plates which were sensible for the green, yellow, and red colours. For as these colouring matters absorb the above-mentioned rays they undergo a chemical change which enables them to decompose the bromide of silver.

The most conspicuous example of the chemical action of the less refrangible rays is, however, afforded by nature herself. Plants draw the whole of the carbon they require for their growth from the air, and this they effect by the decomposition of carbonic acid gas, which they break up into carbon, which remains as part of the plant, and oxygen which is returned to the atmosphere in the gaseous form. This action, so important for the welfare of plants, is completed only in the green (chlorophyll-holding) parts of the plants under the influence of the solar light. By means of researches on different coloured light it is now ascertained that those rays which cause the liveliest elimination of oxygen belong to the less refrangible half of the spectrum.

CHAPTER XIV.

ACTION OF HEAT.

83. THE surface of the earth is not only illuminated by the solar rays, but it is also warmed by them. It is clear from what has been said that rays which are reflected from the surface of any body, or which are transmitted, cannot have any action in warming it. It is by the retained or *absorbed* rays alone that it can be warmed.

From this point of view it is not difficult to apprehend the different behaviour of bodies in regard to their capacity of being warmed by the solar rays.

Air being transparent allows the solar rays to traverse it without diminution of their intensity; it is consequently warmed by them only to a very insignificant degree. Hence the upper regions of the air, although they receive the solar rays at first hand, are so cold that even in the tropics the summits of high mountains are covered with everlasting snow. The warming of the air is mainly due to the heat it receives from the heated surface of the earth below, which gradually communicates the heat it has obtained by absorption to the strata of air in immediate contact with it.

Bodies with polished surfaces, which reflect the greater part of the rays falling upon them, and transparent colourless bodies, which almost wholly transmit such rays, undergo only slight heating. On the contrary, rough surfaces, that is to say, surfaces incapable of much reflexion, and dark colours, or those which possess high power of absorption, are conditions that favour the heating action of light.

Any substance therefore will become heated by radiation to the greatest degree when its surface is made rough and completely black, so that it can absorb all the rays falling upon it. This object is best attained by coating the substance with lampblack.

Thus, for example, if two thermometers be exposed to the sun, the bulb of one of which is blackened whilst the other is bright, the former will show a higher temperature than the latter.

Herschel first suggested that with the aid of such a blackened thermometer the calorific power of the different coloured rays of the spectrum could be tested. When he exposed a thermometer successively to the several rays he found that the red were much hotter than the blue, and that even *in the dark region on the near side of the red end* a considerable elevation of temperature was still observable.

An ordinary thermometer, however, is not sensitive enough to follow and determine all the degrees of variation of temperature in the spectrum. But we possess in the *Thermopile* an instrument admirably adapted for such delicate researches.

If rods of antimony and bismuth be soldered together in the manner shown in fig. 127, so that the first, third, and fifth, &c., or generally the odd numbered joints, are turned in one direction, whilst the even numbered joints are turned to the opposite side, and if the terminal rods a and b are connected by a wire, an electric current is excited in this as soon as one series of joints, as, for example, the odd numbered joints, are heated.

These groups of rods are enclosed in a brass case (fig. 128), so that the odd numbered joints lie between the slit $a\ b$, whilst the terminal rods are connected with the binding screws c and d. The joints are blackened, to favour as far as possible the absorption of the rays falling upon them. This apparatus is termed a *Thermopile*; and because the joints are arranged in a *straight* line, $a\ b$, a *linear* Thermopile.

FIG. 127.

Construction of the thermopile.

FIG. 128.

The strength of the *thermo-electric current* traversing the wire connecting the poles is proportional to the heat applied to the joints. From the intensity of the current may be estimated the degree of heat to which the joints have been exposed.

For the measurement of the intensity of the current the instrument termed the *Galvanometer*, and depicted

Linear thermopile.

in fig. 129, is employed. A copper wire covered with silk is wound round and round a frame of wood, in the interior of which a magnetic needle is freely suspended by means of a fibre of silk from the cocoon. The ends of the wire are fixed by binding-screws. A second magnetic needle, firmly connected with the first, is placed above the frame, and plays freely over a circle divided into degrees. The two needles are parallel to each other, but their poles point in opposite directions. By this means they are retained in the position of equilibrium resulting from the magnetism of the earth with very slight force only, whilst the action of the current, which exerts its influence alike upon both, is doubled. The action of a galvanic current traversing the coil consists in causing the needles to deviate from their position of equilibrium parallel to the turns of the wire, and this to an extent corresponding to the intensity of the current.

Fig. 129.

Galvanometer.

84. If now the binding-screws of the Thermopile are connected by means of wires with the ends of the coil of the Galvanometer, and the Thermopile be placed in the violet end of a solar spectrum thrown by a flint-glass prism, it will be found that the deviation of the galvanometric needle is extremely small; but it will be observed that the deviation progressively increases as the Thermopile is gradually moved towards the red end of the spectrum, and that it even becomes still greater in the dark region on this side of the red till a point is reached which is as distant from the line B as this is from the line D. From this point onwards it gradually again diminishes, though it may be followed for a considerable distance into the dark region.

Thus it is seen that amongst the rays emitted by

ACTION OF HEAT. 201

the sun there are some of still less refrangibility than the extreme red rays, and these may be termed the **ultra**-red rays. They are recognised by their calorific action alone; they are imperceptible to the eye, for the reason that they are absorbed by the fluids of the eye, and never reach the retina.* On this account they are sometimes termed the 'dark calorific rays.'

In order to obtain a general view of the calorific action of the different parts of the spectrum, perpendicular lines must be erected upon the long axis of a spectrum (fig. 130) of a height corresponding to the measured heating power of that part of the spectrum. By joining the apices of these perpendiculars we obtain a curved line which exhibits the varying amount of the calorific power in different parts.

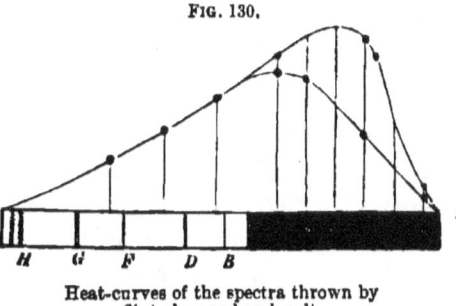

FIG. 130.

Heat-curves of the spectra thrown by flint glass and rock salt.

In the spectrum of a flint-glass prism the apex of the thermotic curve—that is to say, the place of greatest heat-effect—is situated, as is shown above, outside the apparent spectrum in the ultra-red region.

If the spectrum thrown by a prism and a lens of rock salt be now examined, the thermotic action will be found exactly equal in the visible part of the spectrum to that of the corresponding part of a flint prism spectrum; in the ultra-red region, however, the thermotic curve of the rock-salt spectrum rises above that of the

---

* According to the researches of Brücke and Knoblauch.

flint spectrum, and its highest point is still less refracted (fig. 130, upper curve). It appears therefore that flint glass is less diathermanous for the dark heat-rays than rock salt. By experiments—an account of which would lead us too far astray—it may be shown that rock salt allows the dark rays to pass without let or hindrance, whilst most other bodies, even if they happen to be quite transparent for luminous rays, absorb them to a greater or less extent. If it be required therefore to compare the spectra from various sources of light in regard to their thermotic action, the prisms and lenses should be made of rock salt.

We thus find, for example, that the electric light from carbon points is relatively much richer in dark thermotic rays than sunlight. At a point of its ultra-red spectrum which is at the same distance from it as the commencement of the green upon the visible side, the thermotic action is, according to Tyndall, *five times as great* as that of the red rays.

The stronger thermotic action of the ultra-red rays, in comparison with that of the luminous, is strikingly shown by the following experiment:—Two spherical flasks are taken, one of which contains a transparent solution of alum, which permits all visible or luminous rays to pass through it without interruption, whilst it absorbs the invisible thermotic rays. The other is filled with a solution of iodine in carbon bisulphide, which appears black because it is completely opaque for luminous rays; it transmits, on the contrary, the thermotic rays. If the alum flask be

Fig. 131.

Action of the invisible thermotic rays.

placed before the aperture of the electric lamp, it collects, acting like a lens, the luminous rays into a *caustic of dazzling brilliancy*, the heating power of which however is but small, for a pellet of gun-cotton placed in the focus will not explode. The flask containing the black fluid (fig. 131), on the contrary, unites exactly in the same way the dark rays into an invisible focal point, the heat of which not only causes the gun-cotton instantaneously to explode, but even raises a piece of platinum foil to red heat.

85. Every source of light gives off, besides its luminous rays, dark rays of small refrangibility. Hot bodies, on the other hand, not heated sufficiently to glow, emit only dark rays. In the Thermopile we possess a means of demonstrating the presence of such rays and investigating their behaviour. And the results of numerous researches have shown that the dark rays obey the same laws as the bright ones; they undergo reflexion from polished surfaces as from a mirror, whilst they are diffusely reflected from rough surfaces. They course in a straight direction through one and the same medium, but are refracted when they enter another medium, their refrangibility agreeing with that of the ultra-red portion of the spectrum.

A solid body, as for example a platinum wire, which is gradually raised to an intense heat, first emits dark ultra-red rays; as soon as it begins to glow, it emits in addition the extreme red rays. At a bright red heat its spectrum extends as far as F, and at a white heat it gives off all kinds of rays as far as H.

All these facts demonstrate that no other difference exists between the dark heat-rays and the luminous rays than the gradual and progressive increase of refrangibility; the former do not differ from the latter otherwise than the red rays differ from the yellow, or the yellow from the green. The invisibility of the former does not consist in any peculiarity of the rays themselves, but is dependent on the nature of our eyes, the fluids of which are opaque for the ultra-red rays.

The dark rays are perceptible to us only through the sensation of warmth they give to us; the luminous rays, on the contrary, act simultaneously on two organs of sense—upon the nerves of common sensibility or touch as heat, and upon the eye as light. Every ray of light is thus at the same time a ray of heat. We are incapable, for example, of separating the heating effect caused by the yellow light of Sodium from its illuminating power. It gives no rays of such low refrangibility that they produce only the effects of heat, and not of light.

Light and radiant heat are therefore, as effects of one and the same cause, to be distinguished from each other not by any peculiarity of their own, but only by us as different forms of sensation. The same individual ray calls up in us, according to the nerve-path through which the impression it makes is conducted to the seat of our consciousness, sometimes a sensation of light and sometimes of heat, just as a drop of vinegar applied to the tongue tastes sour, but if brought into contact with a sore place on the skin, produces a sensation of burning; or as a tuning fork when struck produces a sensation of sound in the ears, but a feeling of vibration to the hand in contact with it.

'86. If now a general view of the solar spectrum throughout its whole extent be taken, it is seen to be composed of three portions of nearly equal length--

the ultra-red, the luminous, and the ultra-violet portion.

In the figure below (fig. 132) three curved lines are drawn above the spectrum, of which that marked by *III* is the curve that we now know of heat; the curve *II* in like manner expresses the chemical action on a mixture of chlorine and hydrogen and the salts of silver; and the curve *I* gives the brilliancy of the illu-

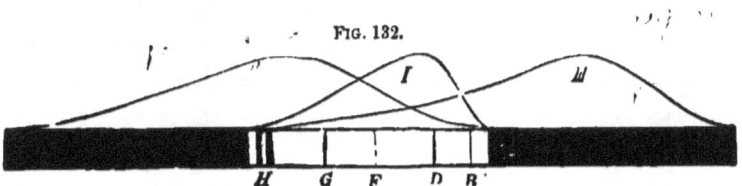

FIG. 132.

Light, heat, and photographic action of the solar spectrum.

mination within the limits of the visible spectrum. From this drawing it is evident that the maximum amount of light is in the yellow, the maximum of the photographic action is in the violet, and finally, the maximum heat is in the ultra-red.

In reference to the rays themselves, these three curves have a very different signification. It is clear that the action which a ray exerts upon a body is determined on the one hand by the intensity or energy of the ray, and on the other by the capacity for absorption of the body. However great the intensity of a ray may be, it will exert no influence upon a body which will not absorb it. Thus, for example, the red rays, however intense they may be, exert no influence on a mixture of hydrogen and chlorine, or sensitive silver salts, because these substances do not absorb them.

Each of the curves *I* and *II* therefore expresses the

co-operation of two actions—the intensity of the rays and the capability of absorption of the retina, or of a photographic plate—which is *very different* for different kinds of rays. They afford us therefore but little direct information on either point. The curve *III* shows the heating influence which each part of the spectrum exerts upon the blackened surface of the Thermopile. Now lampblack behaves as an almost perfectly black body to all kinds of rays alike, since it completely absorbs them all, and becomes heated in proportion to their intensity. The thermotic curve shows therefore the intensity of the radiation which falls on each part of the spectrum free from the influence of any special capacity for absorption. It is therefore to be regarded as the *true curve of intensity of the prismatic spectrum*.

CHAPTER XV.

MIRROR EXPERIMENT OF FRESNEL.
UNDULATORY MOVEMENT.

87. THE reader has hitherto had his attention confined to the experimental investigation of the laws of the phenomena of light without speculating as to what light essentially is. A series of phenomena now present themselves which raise again this question of the nature of light, and at the same time afford the means of answering it. Let two mirrors, AB and BC (fig. 133), be made of black glass and be so placed as to meet at the vertical slit, B, the one, BC, being permanently fixed in a wooden frame (Holzklötzchen) which can be moved along a vertical rod and fastened by a wooden screw T, whilst the other, AB, is revolvable by means of the screw S around the angle B by means of the hinge attached to it. The moveable mirror is to be placed in such a position that

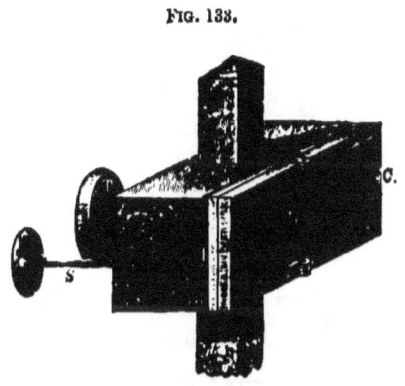

FIG. 133.

Fresnel's mirror.

15

208 OPTICS.

its plane forms a very obtuse angle (not differing much from 180°) with that of the fixed mirror.

A sharply defined point of light is required, and may be obtained by letting the solar rays proceeding from a Heliostat fall upon a lens (fig. 134) of short focal distance, which unites them into a focus P. The luminous point P emits rays which strike both mirrors; from the mirror $A B$ they are so reflected that they

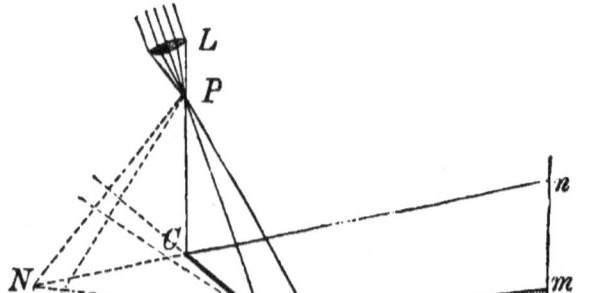

Fig. 134.

Fresnel's mirror experiment.

appear as if they came from the image-point M of this mirror. The mirror $B C$, on the other hand, reflects the rays as if they proceeded from its image-point N. In order that the two mirrors may each have only one reflecting surface and have only one image-point, they must be made of black glass or of metal.

From these mirrors two cones of light $M m m'$ and $N n n'$ are obtained, which appear to proceed from the points M and N. They have the space $B m n$ (shaded in the figure) common to both, so that the field between *m* and *n* upon the screen *m' n'* situated in the path of the reflected ray receives light simultaneously from the two cones of light. In this middle area a series of vertical *dark lines* are perceived, but if one of the glasses be covered the lines immediately vanish and the area which now receives only the light from the opposite mirror appears to be uniformly illuminated throughout its whole extent. The lines however immediately reappear if the cover be removed, and to the light proceeding to the screen from the point *M* is added that also which proceeds from the point *N*.

It has thus been demonstrated that *light added to light may, under certain circumstances, cause darkness.*

If by turning the screw *S* (fig. 133) the angle of the two mirrors be made less obtuse, the lines become narrower and closer together till they ultimately become so fine that they can no longer be distinguished. Hence to render the lines distinctly perceptible the angle between the two mirrors must be very obtuse, or what comes to the same thing, the mirror images *M* and *N* must be very closely approximated.

Instead of making the experiment with a screen so that many can see it at the same time, any individual may observe it directly by making his retina take the place of the screen. This subjective method of observation has the advantage that a feeble source of light may be employed; and then, if the homogeneous light of the Sodium flame be used, the entire field of vision may be observed to be filled with numerous vertical and completely black lines.

88. The just-described mirror experiment of Fresnel, named after the genial physicist who conceived it, teaches that light combined with light may, under

certain circumstances, produce darkness. What then must be understood by the term 'light,' to enable this apparent paradox to be explained?

This much is certain, that every luminous body must be regarded as the seat of a motion which is by some means propagated to our optic nerves and arouses in them the sensation of brightness.

Two modes, however, are only known in which movement may be propagated from one point of space to another.

The first mode is the *immediate transference of motion* in which the moved body itself or parts of the same traverse the space between the two points, as when a cannon ball flies to its goal from the cannon.

The second mode of transference takes place *mediately* through an elastic medium intervening between the two points, in which *medium* the body originally in motion excites a vibratory movement that is propagated from particle to particle, it may be to a great distance, without a particle of the originally moving body itself or any portion of the propagating medium moving from its original position to any considerable extent. This process is called *undulatory movement*.

As an example of the former, the sense of smell may be taken, which is excited by the immediate transference of particles of the odorous material to the olfactory organ. If a flask containing some ammoniacal gas, which is colourless, be opened, those near it quickly perceive the stimulating odour of the gas, whilst it is only perceived by those who are more distant after the lapse of some time. It would be easy to demonstrate by appropriate tests the presence of particles of ammonia even in the furthest corner of a room. The smell is perceived still more strongly if a second flask be opened, so that the number of particles of ammonia present in the air is increased; it would, however, be needless to do this, since all must be satisfied that the sense of smell is excited by particles of the odorous material which come into direct contact with the olfactory organ, and that by increase of the effective particles alone can the intensity of the sensation be augmented.

Another of our senses, hearing, on the other hand, receives its impressions through the second mode of propagation, since every resounding body puts the air around it into undulatory movement. If a bell be struck its sound is heard simultaneously with the blow. The blow makes the bell vibrate, that is to say, causes its particles to make rapid to and fro movements or vibrations which are felt by the hand in contact with it as a trembling. The vibration communicates itself in the first instance to the particles of air in immediate contact with the bell, and as these move to and fro in the same rapid manner they produce the same effect upon the particles of the next adjacent layer of air as the bell itself, and set them in motion. In this way the vibratory movement is propagated with great rapidity from one layer of air to another, and finally, on reaching the ear, excites in the auditory nerve the sensation of sound. But it is certain that neither particles of the bell itself, nor even particles of the air immediately surrounding the bell, penetrate the ear; if they did, as sound travels at the rate of 1,116 feet in the second, they would strike on the tympanum with a velocity exceeding that of the most violent hurricane. An extremely simple experiment may now be considered, which may be shown with two perfectly similar organ-pipes standing on a wind-chest common to both. If each pipe be made to speak separately both will give precisely the same fundamental note. If, now, both pipes be sounded together, exactly the opposite occurs to what might be expected; instead of the fundamental note being increased in intensity it is remarkably weakened, so much so, indeed, that at a little distance from the pipes the fundamental note is no longer audible.

From this circumstance the same conclusion is drawn in regard to sound, which unquestionably consists in an undulatory movement, as was done in the case of the light in the mirror experiment of Fresnel, namely, that *sound added to sound may, under certain circumstances, produce silence.*

89. Through which of the two possible modes of propagation does the movement that we call 'light' spread? Are our eyes when we look at the sun struck by particles of a luminous material uninterruptedly emitted by that luminous body? Or do the rays of light consist of a vibratory movement which strikes upon our retina in the form of minute waves—in other words, is the process of seeing analogous to that of smelling or of hearing?

The choice between these two modes of explaining the phenomena, after what has been said, cannot be difficult. On the supposition of there being a *luminous substance* (emission theory), the fact that light superadded to light can produce darkness is wholly inexplicable. On the other hand, a case has been cited in which an undulatory movement co-operating with a similar undulation produces such an effect, and we shall see immediately that this follows necessarily from the very nature of undulatory movement. It will, moreover, be seen that the admission of *luminous waves* (undulatory theory) gives a perfectly satisfactory explanation, not only of the phenomena in question, but of the great majority of the phenomena of light, and is opposed to none of them, whilst the conception of a luminous æther or substance has long been negatived by facts.

As the view that light is itself a material substance is set aside, and it is regarded as an undulatory movement, it becomes necessary to admit the existence of a material in which the waves of light can propagate themselves. The air, in which the waves of sound spread, cannot be coincidently the carrier of luminous waves, for it only forms a thin investment around our earth, and perhaps other heavenly bodies; whilst in the immeasurable depths of space through which the light of the sun and the fixed stars reaches us no air is present. It must therefore be admitted that the universe is filled with an elastic material which is so rarefied that it opposes no appreciable resistance to the movement of the celestial bodies. This attenuated elastic matter is called 'Æther.'

90. The waves of water afford an excellent representation of the phenomena of wave-movement. If a stone be thrown into water at rest a circular depression forms around the point struck which spreads wider and wider with uniform velocity. In the meanwhile an elevation has formed at the point where the stone entering the water had originally caused a depression; then as this sinks back to its original level it produces a wall-like circular elevation around it, which follows

up the preceding circular depression with equal velocity. Whilst the fluid continues its up-and-down movement at the point struck, fresh alternately depressed and elevated wave rings appear to proceed from this middle point, or, as it is customary to call them, wave elevations (crests) and depressions (sinuses) (Wellenthäler and Wellenberge), are formed, which, owing to their constantly spreading more and more widely give the illusory appearance of the fluid streaming out on all sides from the middle point.

That no such streaming movement does really occur may easily be demonstrated by observing any small object accidentally floating on the water, as for example, a piece of wood. This, as the crests and sinuses of the waves spread beneath it, merely rises and falls without materially changing its original position, making the oscillation of the particles of water immediately beneath it apparent.

The cause of the waves of water is the force of gravity which is exerted after each disturbance of the equilibrium to restore the fluid to its original horizontal plane. Whilst the particles of the water first struck and depressed by the stone are soon again compelled to rise, they oblige at the same time the easily moveable adjoining particles to descend in order that the depression which was formed may be again filled up. As every particle begins to fall somewhat later than the immediately antecedent one, a circular wave-depression spreads round the central point of excitation, which attains its full development at the moment in which the particle struck in its ascending movement has again attained its original level. It does not however here come to rest, but continues its movement upwards above the horizontal plane of the water until the force of gravity acting in opposition has exhausted its upward directed velocity, and it swings back again to the level. In the meanwhile the neighbouring particles, which exactly imitate the undulating movement of the first disturbed particles in the same period of time, form a wave-crest which is fully developed at the moment in which the first particles have again reached the plane in their descending movement.

And now, when the particle first excited has completed *one entire vibration*, and is, as at the commencement of its movement, again about to leave its position of equilibrium in order to descend, it has around it a *complete wave*, consisting of a wave depression and a wave crest. This wave as it spreads produces the second to-and-fro movement of each particle, and every subsequent complete wave acts in a similar manner, and as the new waves immediately follow those antecedent to them, a circular system of waves is developed around the central point of excitation.

91. Every straight line drawn from the middle point of a system of waves upon the surface of the water regarded as horizontal is termed a *wave ray*. All particles of water which when at rest lie on this straight line are now elevated, now depressed, according to whether they for the moment belong to a wave crest or a wave depression, and form therefore in their serial succession an ascending and descending sinuous line. Such a *wave-line*, granting that the particles rise and fall perpendicularly to the ray AB, is represented in fig. 135. That portion of a ray which is included in a complete wave, that is to say, which includes a wave crest and a wave depression, or any portion of it equal

to this, is called a *wave-length*. In the figure we have for example between A and B three complete wave-lengths, and one wave-length between b and c, and between c and d. Those particles which in any ray are separated from one another *one or several complete wave-lengths,*

FIG. 135.

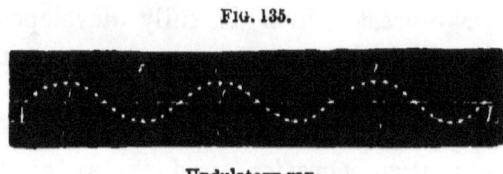

Undulatory ray.

are at any moment of time in exactly the same condition of undulation, their movements are in perfect accordance with each other. The particles b' c' and d', for example, which are distant from one another one or two wave-lengths, have all three arrived at their greatest height, and are about to descend. Moreover, the particles A and B, the distance between which includes three wave-lengths, are both in the act of descending through their position of equilibrium.

The particles b' and f' on the other hand, which are distant from each other a half wave-length, are in just the opposite conditions of vibration. For whilst the former is beginning to fall from its highest position, the latter is just about to rise from its lowest position. The same relation occurs between the particles f'' and d', which are distant from each other three half wave-lengths. Speaking generally, it is clear that the movements of two particles the *distance between which is an unequal multiple of a half wave-length are directly opposed.*

CHAPTER XVI.

PRINCIPLE OF INTERFERENCE. CONSEQUENCES OF FRESNEL'S EXPERIMENT.

92. WHAT happens if *two* wave-systems meet on the same fluid surface?

If from a vessel held above a flat pan containing mercury two fine streams of mercury are allowed to fall, each produces around the point where it strikes the surface of the fluid a circular system of waves. As the two wave-systems decussate they divide the surface into a regular network of small elevations and depressions, a representation of which is attempted in fig. 136.

If the light of the sun or of the electric lamp be allowed to fall upon the surface of the mercury, the reflexion upon a screen will also furnish a representation of this delicate phenomenon.

It is not difficult to explain the effects observed. At all points where two wave crests meet, the surface of the fluid, if the two waves are equal, rises to twice the height, and where two depressions meet it sinks to double the depth. At those points on the contrary where a wave crest is cut by a sinus, the upheaving and depressing forces are in equilibrium, and the fluid remains at rest at its original level.

In a fluid set in motion by two or more equal or unequal wave systems, every particle, speaking generally, undergoes a change of place, which is the sum of all the movements impressed upon it by the several systems of waves at the same moment. Of course, by the

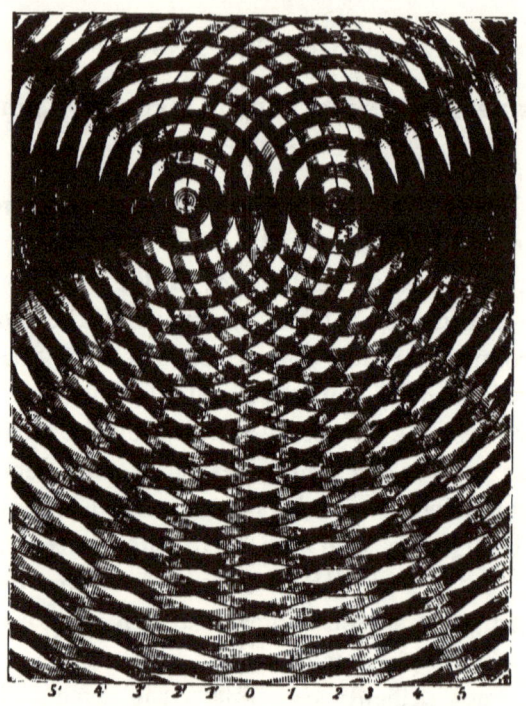

Interference of two systems of waves.

word 'sum' the so-called *algebraic sum* is meant, that is to say, the elevations are regarded as positive, the depressions as negative values.

In other words, it may be said that every wave system superimposes itself upon, or adds itself to, a surface already moved by waves, as it would do were it acting alone on the surface at rest. Every wave system forms itself unhindered by those already present, and spreads after it has crossed these upon the still quiescent surface of the water as if it had suffered no interruption. We see, for example, the slight wave rings excited by the falling rain drops form on the larger waves raised by a steamboat just as well as upon the sea at rest. It may be observed again that these waves, when they traverse an area rippled by the breeze, take the small waves on their back, and having passed beyond this region leave these last behind with their original form unaltered.

The important law just laid down, to which the processes taking place in the co-operation or *interference* of two or several systems of waves are subjected, is termed ' *the principle of interference.*'

93. Returning to the simplest case of interference of two equal systems of waves represented in fig. 136, it appears that an explanation can be given of the movement occurring at each point of the surface of the fluid, if, instead of the waves themselves, the *wave rays* are kept in view. If we consider, for example, the points 5 .... 5' lying along the wall of the vessel, the two rays which may be conceived as drawn from the two middle points of the exciting cause of them to the central point O are equal to each other in length; the oscillating movements which proceed simultaneously from each of these centres meet therefore in the point Q under equal conditions and produce the greatest possible effect. In the laterally situated point 1, on the other hand, two rays meet which are about half a wave different; the forces which they exert upon the point are therefore equal and opposite; the point consequently remains at rest. The same occurs at 3 and 5, where the difference between the rays cor-

responds respectively to 3 half and 5 half wave-lengths. At the points 2 and 4, on the contrary, where the rays respectively differ one or two entire wave-lengths, and thus meet under equal conditions of oscillation, the liveliest movement takes place. The intervening points are maintained in less active movement by pairs of rays of all possible degrees of accordance and opposition.

The points 1, 3, 5, ... 1′, 3′, 5′ .... thus remain at rest under the action of the two systems of waves. That which in waves of fluid is rest, is in waves of sound silence, and in waves of light darkness.

It is scarcely necessary to expressly mention here that this affords a complete explanation of Fresnel's mirror experiment, and that fig. 136 is a sketch of it. If, for example, the two points of light produced by the mirrors M and N (fig. 134) be regarded as centres of origin of light waves, and the wall 5′ .... 5 as the screen for receiving them; and if it be further considered that the waves of light expand, not only circularly in one plane, but like a sphere into the surrounding æther, it will be understood that, in consequence of the interference of the two systems of waves, vertical dark lines must appear at the points 1, 3, 5 .... 1′, 3′, 5′, and bright striæ at the points 2, 4 .... 2′, 4′.

But why, it may perhaps be now asked, should the two points of light be employed in a roundabout way after their reflexion in the two mirrors? Would it not be simpler to put aside the mirrors, and use, instead of the images M and N thrown by them, two luminous points like the points of a glowing platinum wire? The answer to this question is obtained from the fact that the two wave systems, in order that they should produce dark lines in the given points of the screen, must proceed simultaneously, and in a precisely similar manner, from the two luminous points. But we are unable so to conduct the process of light production in two luminous bodies, or even in two points of a single luminous body, as to make the undulating movement proceeding from one exactly accordant with that of the other; in each of them, after a short period, interruption of the movement, augmentation and diminution of the liveliness of the flame, and other disturbances take place, which do not occur coincidently in the other. Hence the lines of interference are only partially formed, and in rapidly changing parts of the screen giving to the eye the impression that it is equally and uniformly illuminated. Two independent and separate luminous points therefore, on account of the inequality of these wave systems, present no interference lines. The equality required for this purpose is obtained with the greatest certainty by making the two wave systems spring by mirrors or by any other appropriate means from the same source. The irregularities to which the process of light production is subjected, whatever may be the light used, take place concordantly and simultaneously in both systems of waves, and consequently exercise no influence upon the accordance and opposition of the rays which are now conditioned only by their difference of path.

94. Fresnel's experiment may now be repeated, with this difference, that a red and a blue glass are placed alternately before the aperture of the Heliostat. It is then seen that with blue light the lines are closer together than in the red, that is to say, the corresponding series of dark lines are in the former case nearer to

the middle bright lines than in the latter. Two blue rays thus require, in order that they may neutralise each other, a smaller linear difference than two red ones; *the wave-length of blue light is consequently smaller than of red light.*

If as brilliant a spectrum as possible be produced by means of a prism, and its coloured rays be allowed to fall successively upon the lens L (fig. 134), and consequently on the mirror, we find that the distance between the lines, and consequently the wave-lengths, become progressively smaller in passing from the red to the violet. This affords an explanation of the reason why, when white light is employed, the lines are not alternately black and white, but coloured. The middle bright lines, in which all colours are mingled in their highest intensity, are completely white, but towards the sides first the violet fades out, and then in succession the several colours from the most towards the least refrangible. The consequence of this is that the middle bright lines towards the interior are edged with yellow, and towards the exterior with red; at the point where the brightest colour, yellow, disappears, the first dark line is seen, which, however, since the violet has here again become stronger, exhibits a faint violet tint. Then follow white, yellowish-red, violet, to the second dark line, which is blue. Then come green, yellow, red, bluish-green; still further on a few alternations of red and bluish-green occur, and very soon, inasmuch as the lines of various colours mingle, only a uniform white remains. White light therefore gives only a few lines, which as we pass outwards constantly become more and more indistinct; when homogeneous light is used, on the other hand, the dark lines are completely black, and are present in great numbers.

Fresnel's experiment, however, not only shows broadly that there are differences in the lengths of the waves, but it enables us to measure these differences.

If, for example, the length of the rays proceeding from the luminous points towards the first black lines be obtained, which can be done with sufficient accuracy, their difference must be equal to half the wave-lengths of the homogeneous light employed. In the lines of higher order which correspond to the differences of path of 3, 5, 7, etc. half wave-lengths, the measurement can be repeated with the accuracy required. Fresnel made these measurements for light which had traversed red glass, and found the wave-length of this red light equal to 638 millionths of a millimeter.

A method will hereafter be shown by which the wave-lengths may be determined with still greater accuracy and for definite rays (for the Fraunhofer's lines). A conception of the extraordinary smallness of the waves of light may be obtained from the statement that in the length of one millimeter there are 1,315 waves of the extreme red (line A), 1,698 waves of the yellow light of Sodium (line D), and 2,542 waves of the extreme violet (line H_2).

95. It is well known that if the performance of a piece of music be listened to at various distances, the same accordance in the notes, the same harmony, is always perceived; the high and the deep notes which fall in the same beat reach our ears in all cases simultaneously. The conclusion from this is that all tones, whether high or low, strong or feeble, *are propagated through the air with equal rapidity*. The rapidity of propagation of

sound, that is, the distance to which the vibratory movement of a resounding body spreads in the air in a second, is estimated at 340 mètres (1115·4 feet).

But, as has been already shown, a complete wave originates with each entire vibration; every sounding body will therefore produce as many successive sound waves in a second as the number of its vibrations in a second, and since the sound in this period of time has spread over a distance of 340 mètres, the total length of the sound waves excited in one second must amount to 340 mètres. The wave-length of a tone is consequently obtained by dividing the rapidity of propagation (340m.) by the number of its vibrations. The wave-length of the tone of an A tuning fork, which makes 440 vibrations in a second, is thus found to be equal to 773 millimeters. The wave-length of every movement the rapidity of propagation of which is known to us, may in this way be deduced from the number of vibrations, and of course also, conversely, from the wave-length the number of vibrations.

The rapidity of propagation of light is so enormously great that even at a distance of 60 miles, which is as far as terrestrial signals will reach, no difference of time can be observed between the moment of emission and of arrival. The velocity of light has, however, been measured by means of astronomical observations, and more recently by physical experiments. An account of the ingenious methods by which this has been accomplished cannot be here appropriately introduced. It is only requisite to state that the concordant results of all measurements show that the light both of celestial bodies as well as that proceeding from terrestrial sources, traverses a distance of about 186,000 miles a second. Some observations by Arago,* and especially also reasons which are theoretically deduced from the nature of undulatory movements, justify us in admitting that *the rapidity of propagation of every kind of light, whatever may be its colour and brightness, is, in the free æther of the universe, alike.*

The wave-lengths of the homogeneous kinds of light, as well as their rapidity of propagation, being now known, the number of their vibrations can be determined with facility. This is expressed by the number of wave-lengths which are contained in the length of 186,000 miles. The extreme red line (A), 1,315 of the waves of which occur in a millimeter, are thus found to have the prodigious number of 394,500000,000000, or in round numbers, 395 billions of vibrations in a second. The shorter the wave-length the greater must be the number of vibrations; in a ray of yellow Sodium light every particle of æther makes 509 billions of vibrations in a second, and the extreme violet line (H_2) corresponds to a number of vibrations amounting to 763 billions.

A musical note appears to our ears higher in pitch the greater the number of its vibrations in a given time; and just as the ear perceives the rapidity of the vibra-

---

* If light of different colours travelled with different velocity, a white star which became suddenly visible would be seen by an observer at a distance of that colour in the first instance which propagates itself with the greatest velocity, and then of a succession of mixed colours till it by degrees became white. If it then again disappeared it would pass through a similar series of colours in inverse order till it dissolved into the slowest-moving colour. Similar phenomena would be exhibited by the variable stars, especially if their period were short, and there were a considerable difference between their greatest and least brightness. Arago undertook a series of observations in regard to Algol in Perseus, which fulfils these conditions, but could perceive no change of colour.

tions of sound as pitch of sound, so does the eye perceive the frequency of the undulations of light as *colour*. Thus for the sensation of yellow characterising the Sodium flame to be produced in our minds 509 billions of æther, neither more nor less, must enter the eye and strike the retina. *Speaking generally, the colour of every homogeneous ray of light is determined exclusively by the number of its vibrations*; the number of vibrations is the objective characteristic of that which we perceive subjectively as colour. The succession of colours in the spectrum is consequently to be regarded as a scale which rises from the lowest tone perceptible to our eye, the extreme red, to the highest, the extreme violet. Antecedent to the commencement of the visible scale in the red, are the deeper ultra-red tones, the vibrations of which are too slow to excite the sensation of light in our optic nerves, and at the other extremity are to be added on as highest tones the ultra-violet which produce only an extremely feeble impression of light in our eyes.

96. It is now requisite that close attention should be paid to a chain of reasoning that will here be offered in regard to a few experiments of the simplest kind.

A close wound spiral coil which hangs vertically in front of a scale divided into centimeters carries at its lower end a plain brass ball. The lowest part of the ball has a little hook. On attaching to this a weight of 100 grammes the elastic coil at once becomes elongated and the brass ball descends two centimeters. With a weight of 200 grammes the elongation is twice as much, or four centimeters, and three times the weight again produces three times the amount of elongation.

Thus it appears that the force which must be applied

to move the ball from its original position in opposition to the elasticity of the wire increases in the same ratio as the amount of displacement effected.

Let the weights be now removed and when the ball has returned to its original position, let it be pressed down with the fingers about two centimeters; then inasmuch as it is kept in this position, the pressure downwards exerted must be identical with the weight of 100 grammes, which was before necessary to effect this elongation, and when the ball is set free it returns with this force to its position of equilibrium.

When, however, it has reached the position of equilibrium it does not at once come to rest, but continues to perform upward and downward movements which are slow enough to permit them to be counted. If the ball be now depressed to the extent of 4 centimeters, and then be set at liberty, it has twice as far to go from its extreme point to the position of equilibrium, or the extent (or amplitude) of its vibration is now doubled. If its vibrations are now counted *the same number of vibrations* will be found as in the former case, for since not only the space traversed but also the expression of force of the tense spiral wire has now been doubled, the greater space must be traversed in the same time. Nor is any alteration observable in the number of vibrations when the ball is drawn down to the extent of 6 centimeters from its position of rest, although the amplitude of its vibration is increased threefold.

From this it appears that the *number of vibrations* is dependent exclusively upon the nature of the vibrating body—upon its internal forces, if we may so speak, —but in no way upon the amount of the *external force* applied to it; the amount of force applied to it finds its expression in the *amplitude of the vibration*. When the ball is depressed four centimeters, the hand has not only to exercise twice as much force, but it has to traverse twice the distance that it has when it is only depressed two centimeters. The *work* which must be performed to overcome the elastic force of the wire in the former case is therefore *four times* as great as in the latter, and if with three times the force the ball be moved over three times the space, nine times the amount of force used in the first instance has to be applied. When the hand is removed the work performed by it is transferred to the ball, and expresses itself in the energy of its vibrating movements. By virtue of this energy the ball, until it comes to rest, performs the same amount of work which was applied to it to set it in movement.

From these considerations it results that the energy of the vibrating movement is proportional to the square of the amplitude of the vibrations.

The facts taught by the vibrating ball are applicable alike to the vibrations of sound and of light. The *tint of colour is dependent on the frequency; the intensity (or energy) of light on the liveliness of the vibrations.* Whilst the former depend on the number of the vibrations, the latter are measured by the square of the amplitude of the vibrations.

CHAPTER XVII.

HUYGHENS' PRINCIPLE.

97. 'LIGHT consists of a very minute vibrating movement of an elastic medium, which is propagated with great rapidity, but not instantaneously, in straight lines that proceed like the radii of a sphere from a central point common to all.'

Hooke,* the accomplished friend and countryman of Newton, who wrote the date 1664 under these words, may be regarded as the first who clearly seized and expressed the fundamental idea of the doctrine of luminous waves. Nevertheless he did not advance so far as to explain the refraction of light by undulatory movement; and he failed because this fundamental idea, in order to be applicable to all the phenomena of light, required still a very important addition to complete and perfect it. It was reserved for Hooke's genial contemporary, Huyghens,† to fill this hiatus, and to become the real founder of the undulatory theory of light.

The theory of Huyghens, so named to do honour to its discoverer, is in fact the egg of Columbus, a simple solution of many complex and enigmatical phenomena, and whilst an attempt is here made to render it intel-

---

* *Micrographia*, Observat. ix.
† *Tractatus de Lumine*, 1690.

ligible, no very great strain will be exerted on ordinary powers of imagination. When an undulatory movement propagates itself through an elastic medium, every particle imitates the movement of the particle first excited. But every particle stands in regard to the adjoining ones in exactly the same relation that the first particle did to its neighbours, and consequently must exert upon those that surround it exactly the same influence as the first. *Every vibrating particle is therefore to be regarded as if it were the originally excited particle of a wave system; and as the innumerable and simultaneous 'elementary' wave systems co-operate with one another at each instant in accordance with the principle of interference, we obtain exactly that 'principal wave system' by which the elastic medium appears at any moment to be moved.*

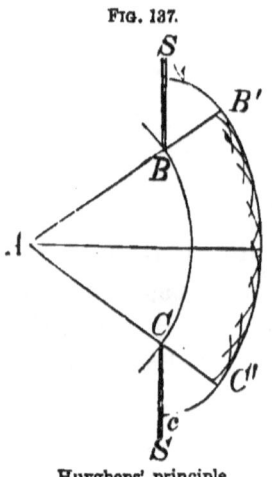

FIG. 137.

Huyghens' principle.

If, for example, all points of the circular or spherical wave BC (fig. 137) which take origin from the centre of disturbance A be regarded as new centres of disturbance, after a little while an innumerable series of elementary waves of equal size will have formed around them, which are represented in the figure by small arcs. The circle $B'C'$ described around the centre A, which all the elementary waves touch at their most distant point, represents the extreme limits to which the undulatory movement has in the meanwhile been propagated. The state of oscillation which previously affected the wave BC is now transferred to the circle $B'\,C''$, to which all elementary waves reach with equal conditions of oscillation. The wave BC has thus propagated itself by means of the elementary waves in the same form and with the same rapidity to $B'\,C''$, as if it proceeded directly from the original point of disturbance A.

The same result is thus obtained whether we admit a direct propagation of a single wave centre outwards, or an indirect propagation effected by innumerable elementary waves. Nevertheless the two modes of explanation are *essentially* different, and the latter is alone *true to nature*, for it alone gives the requisite consideration to the various relations that occur between the particles of an elastic medium. The former more simple mode of explanation may, however, be admitted if, as in the preceding Chapter, we are dealing with those characters of wave movement which are common to both methods of propagation. As long as a wave movement is propagated without disturbance, the elementary waves withdraw themselves from observation because they proceed by their co-operation to produce the chief waves. They immediately appear independently, however, if their adjoining waves are in any way suppressed. If, for example (in fig. 137), the wave BC proceeding from A passes through the opening BC of a screen, it continues its course undisturbed between the two marginal rays AB and AC, whilst the elementary waves proceeding from their points between B and C combine in the manner above described to form the chief wave $B'\,C'$. The elementary waves $B'\,b$ and $C'\,c$ proceeding from the marginal points B and C remain partially isolated, and transfer a movement which, in comparison with the main wave, is, as may be supposed, very feeble, into those lateral spaces $B'BS$ and $C'CS$, which are protected from the main wave.

Phenomena of light will soon be referred to which are caused by a similar lateral expansion of elementary waves. Some further consideration must however still be given to the behaviour of the principal wave.

98. The chief wave proceeding from the combined action of the elementary waves spreads itself, as has been seen, around a luminous point in a concave sphere just as if the propagation took place directly from this point. Both modes of explanation permit us equally to explain the movement of light as a rectilinear radiation from a centre. Careful consideration however shows that there is an essential difference between the two views. Whilst, on the older theory, a direct propagation along a single straight line, that is, the possibility of an isolated ray of light, was accepted; on the other theory in view of the action which every particle of æther exercises upon the adjoining ones, *the existence of an isolated ray of light is inconceivable.*

Nevertheless a ray of light may be conceived as the expression of the *direction* in which the small portion of wave belonging to it lying upon the surface of the sphere is propagated. Speaking generally, the wave itself or parts of the same, must constantly be kept in view if it be desired to draw any conclusions on the laws of the phenomena of light.

However small a portion of the wave surface may be represented, it contains innumerable rays, which collectively form a *beam*, or fasciculus of rays (Strahlenbündel). In point of fact, in optical experiments individual rays of light are never dealt with, but always beams.

The statements formerly made on the supposition of the existence of individual rays of light, however, lose none of their force through the different conception just gained. They still remain perfectly accurate, even when each 'ray of light' is regarded as only the representative of the very thin beam to which it belongs.

In free æther, as well as generally in every medium in which the undulations of light propagate themselves spherically with equal velocity, every ray is a radius perpendicular to the wave segment corresponding to it. If we imagine the wave segment to be very small or very remote from the centre of the sphere, the perpendicular rays falling upon it may be regarded as parallel to each other, and the wave segments themselves as plane. Speaking generally, every fasciculus of parallel rays is propagated by plane waves which are perpendicular to the direction of the radiation.

99. Now that by means of Huyghens' theory we have given an explanation of the real mechanism of undulatory movement, we shall proceed to inquire what happens when a wave of light reaches the plane surface of junction of two different media, as for example a surface of water at rest.

In fig. 138 ab represents a plane portion of an undulation, and $A a B b'$ the parallel fasciculus of rays corresponding to it. As the wave moves onwards towards the surface MN, the particles of æther between a and b' gradually become affected by the movement; every point struck becomes, in accordance with the theory of Huyghens, itself a centre of disturbance, and sends forth an elementary wave into the first medium (the air) as well as into the second.

Let us now consider in the next place the elementary waves returning into the first medium.

At the instant at which the point b' is reached by the undulatory movement, an elementary wave has formed around the point originally disturbed, a, the radius of which must be equal to the line $b\,b'$, to which extent the chief or principal wave has in the meanwhile progressed; this elementary wave is indicated in the figure at c by an arc described from the point a. In like manner the points lying between a and b' have produced ele-

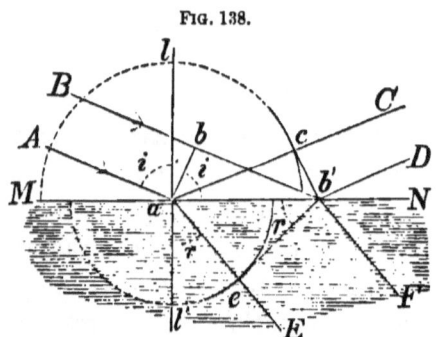

Fig. 138.

Explanation of reflexion and refraction.

mentary waves the radii of which are smaller in proportion as they are nearer to the point b' which is still at rest. If from b' the tangent $b'c$ be drawn to the first elementary wave, it touches also all the other elementary waves, and consequently represents the principal wave which results from the co-operation of all the elementary waves. This wave $b'c$, which is reflected in the direction of the fasciculus $a\,C\,b'\,D$ into the first medium, forms with MN the angle $c\,b'a$, which is obviously equal to the angle $b\,a\,b'$ of the 'incident' wave. If at a upon the limiting plane MN the perpendicular $a\,l$, the 'perpendicular of incidence,' be raised, the angle $A\,a\,l$, which the incident ray $A\,a$ forms with the same, is equal to the angle $b\,a\,b'$ which the wave corresponding to it forms with $M\,N$, and the same holds good for the reflected ray $a\,C$. *The angle of reflexion is thus always equal to the angle of incidence.* We see therefore that the law of reflexion is a necessary consequence of the undulatory theory.

But elementary waves also penetrate into the second medium from the point of the surface disturbed, though the rapidity of propagation is different from that in the first medium. The elementary wave proceeding from the point a must therefore at the instant in which the incident wave reaches the point b', possess a radius $a\,e$ which stands in the same relation to the radius $a\,c$ ($= b\,b'$) of the wave reflected from this point into the first medium, that the rapidity of propagation of the light in the second does to that in the first medium. In the figure $a\,c$ is smaller than $b\,b'$, that is, the rapidity in the second medium is taken as being smaller than in the first. As the tangent $b'\,e$ drawn from b' to this first elementary wave touches also all the other hitherto formed elementary waves, and consequently includes these movements in itself, it represents the plane principal wave penetrating into the second medium. The direction of the fasciculus $a\,E\,b'\,F$ corresponding to it is given by the line $a\,e$, which is drawn from a towards the point of contact E. It is now plain that the ray $a\,E$ forms an angle r with the perpendicular $a\,l'$, which differs from the angle of incidence i, and in our case is *smaller* than this. The ray $A\,a$ has consequently experienced a deflection from the perpendicular in its passage from the first into the second medium. The refracted wave $b'\,e$ forms the same angle r with the surface $M\,N$.

If the particular line here shown, $a\,b'$, be now taken as unity, $b\,b'$ is the sine of the angle of incidence i, and $a\,e$ the sine of the angle of refraction r. The length of the line $b\,b'$ stands in the same ratio to that of $a\,e$ as the rapidity of the propagation of light in the first to that in the second medium. This relation is invariable, whatever may be the magnitude of the angle of incidence. We thus arrive at the proposition that

The sine of the angle of incidence holds an invariable and unalterable ratio to the angle of refraction.

The foregoing statements have, however, not only shown that the law of refraction is a necessary consequence of the undulatory theory, but they also supply a key to the proper signification of this unchangeable proportion which we have hitherto designated as the 'index of refraction.' *The index of refraction in the passage of light from one medium into another must be equal to the relation that the rapidity of propagation of light in the first medium bears to its rapidity in the second.*

As the index of refraction from air into water is equal to $\frac{4}{3}$, the velocity of light in air as compared with water must be as $4 : 3$. If in the former it amount to 300,000 kilometers (186,414 miles), it must be one-fourth less in water, namely 225,000 kilometers (139,810 miles). Speaking generally, it is a necessary consequence of the undulatory theory, *that light is propagated more slowly through strongly refracting than in feebly refracting media.*

It is necessary here to revert for a moment to the view, that light is a peculiar kind of matter, notwithstanding that this view has been on good grounds set aside. On this view refraction is explained by supposing that the particles of the refracting medium exert an attraction or influence upon the particles of the supposed luminous substance, and the conclusion is arrived at that light propagates itself *more rapidly* in the strongly refracting medium than in the feebler one. The direct contradiction which is presented by these opposite conclusions affords an opportunity of finally settling the long contest between the material and undulatory theories of light. Foucault has shown by means of very ingenious experiments that light does travel more slowly in water than in air. If therefore the reasons formerly adduced should still be considered to leave any doubt in regard to the nature of light, there can now be no question that the undulatory theory must be regarded as the only true theory of light.

100. Before proceeding further an attempt must be made to remove another doubt which might arise in regard to the considerations from which the law of reflexion, as well as that of refraction, have followed. It might be objected, namely, that these considerations should be applicable if the *same* substance existed below the limiting surface MN as above; and that we should then obtain instead of the refracted ray, the rectilinear continuation of the incident, but also always reflected ray.

Since now in this case the position of the plane MN could be imagined anywhere, it would result, in opposition to facts, that in a medium of uniform nature light could not only propagate itself forwards from the source of light, but also from all points backwards, and consequently even backwards against the source of light.

That the rapidity of propagation of light in a refracting medium is smaller than in the surrounding air may be explained on the not improbable assumption that the æther contained in a solid or fluid body possesses a *greater density* than that contained in the air, or the free æther of space.

In order to give some idea of what happens when an undulatory movement arrives at the limiting line of two media of dissimilar density, an analogy may be employed. If two ivory balls (fig. 139) be taken, of unequal size, hanging by threads and in contact with each other, and the smaller ball be raised and allowed to fall against the larger, the latter is set in motion in the direction of the blow; the smaller one, on the contrary, rebounds and moves in the opposite direction to that which it originally had. After both balls have again come to rest, if the larger ball be raised and allowed to strike the smaller one, it will be seen that whilst this is driven forwards the former still moves, though more slowly, forwards. In both cases then the striking ball, after it has parted with a portion of its motion, still continues to move.

Fig. 139.

Impact of elastic balls.

Not so if two balls of equal size (fig. 139) be made to strike one another. The striking ball now remains at rest whilst it transfers the whole of its motion to the ball struck, compelling this to move onwards.

The transference of motion from a vibrating layer of æther to a quiescent one, i.e., the propagation of light, is performed under precisely similar laws. If both layers are of equal density, and hence of equal mass, the second acquires the entire motion of the first, which itself remains at rest until it again receives a new impulse from behind, that is to say, from the source of light. In one and the same medium therefore no backward-moving elementary waves can arise. But if the density, and consequently the mass, of the æther layer struck be greater or smaller than that of the striking layer, the latter retains a portion of its motion and gives rise to those backward-going elementary waves which combine to form a reflected principal wave. From this it is seen that reflexion can only occur at the limiting surface of two layers of æther of unequal density.

101. The experiment with unequal balls attracts attention to another circumstance which accompanies the reflexion of luminous waves, namely, that the striking ball maintains the direction of its movement, or the reverse direction, according to whether its mass is greater or smaller than that of the ball struck. In like manner the undulatory motion which is retained by the last layer of the first medium and produces the reflected wave, moves in the same or in the opposite direction to the movement of the incident ray, according to whether the first medium is more or less dense than the second. In reflexion at a denser medium the undulations of the reflected ray are directly opposite to those which it would make were it the immediate continuation of the corresponding incident ray. But we now know that in any ray, those particles that are distant from each other a half wave-length are in opposite conditions of movement. If we therefore imagine the wave line (fig. 135) to be pushed back half a wave-length, the motion of all the particles becomes reversed. The

result previously obtained may therefore be expressed in the following way:—

In reflexion at the surface of a denser medium the reflected ray undergoes a retardation in respect to the incident ray of a half wave-length. In reflexion at the surface of a less dense medium, on the other hand, no such retardation occurs.

102. The velocity of light being smaller in a refracting medium than in air, a ray of light traversing a glass plate, for example, must experience a *retardation* in comparison with a ray of light which has travelled the same distance in air, and this retardation will be greater in proportion to the thickness of glass traversed.

If we apply this consideration to the rays which emanating from a luminous point strike upon a convex lens and unite on the other side into a focus, it may appear at first sight as if, because they have to traverse very different paths, they must strike it with very different velocities. But it is not so. They arrive at the focal point with equal conditions of undulation just as if they had all traversed the same path. If we compare any given lateral ray with the axial ray, the former has indeed a longer path to traverse in the air, but a consequently less thickness of glass; and a more exact examination shows that the greater retardation which it experiences in the air is completely compensated for by the smaller retardation in the glass.

A lens therefore produces no difference of velocity in the several rays of a fasciculus, inasmuch as it unites them in a (real or virtual) focus, and allows those differences which were originally present to remain unaltered.

This is the reason why we observe the lines of

Fresnel and other phenomena of interference subjectively, that is to say, through the eye (which is indeed nothing but an apparatus of lenses) either with or without a lens or telescope, without any disturbance of the phenomenon through the action of the lens.

CHAPTER XVIII.

DISPERSION OF LIGHT. ABSORPTION.

103. AFTER having acquired a knowledge of the true significance of the index of refraction as the relation of velocity of propagation in the first to that in the second medium, it is easy to express the facts of the dispersion of colour in the language of the undulatory theory. To say that waves of different colour undergo an unequal amount of refraction is equivalent to stating that *in a colour-dispersing medium the various homogeneous kinds of light are propagated with different velocities.*

The proposition, that *all kinds of light are propagated with equal rapidity,* which we were formerly compelled to admit in regard to the free æther of the universe, is thus no longer admissible for the æther contained in the interior and occupying the interstices of the particles of natural substances.

The action which the particles of a body exert upon the undulations of æther propagating themselves in it, may be conceived to be dependent on the nature of these particles. In very many fluids and solids, especially in the colourless and transparent ones, as water, glass, &c., the rays produced by more rapid undulations are more strongly deflected, that is to say, experience a greater amount of retardation than the rays produced by a smaller number of undulations. Prisms composed of such substances exhibit a spectrum with the ordinary succession of colours, from the least refrangible red to the most refrangible violet rays. The specific nature of the substance is however rendered evident even here by the different arrangement of the lines of Fraunhofer. (*See* fig. 106.)

The dispersion of colour in atmospheric air and in gaseous bodies generally is (according to Ketteler) so insignificant that we may admit in them, as in free æther, equal velocity for all kinds of light, being smaller than that of universal space in the proportion of 1 : 1·000294. This number expresses the index of refraction of a ray of light in its passage from empty space, that is to say, space filled with free æther alone, into air at a temperature of 0° C. and under 760 millimeters pressure.

The influence of the nature of the material particles on the velocity of propagation is remarkably

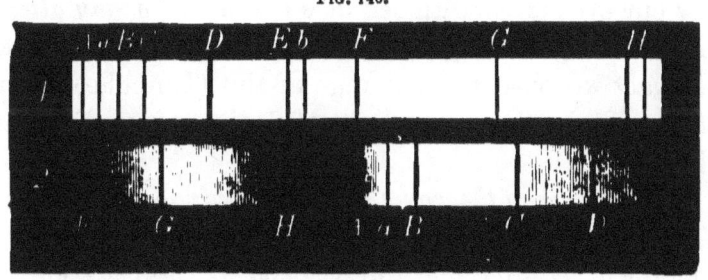

Fig. 140.

Unusual dispersion power of Fuchsin.

exhibited in coloured substances, especially in those in whose absorption spectra one or more very dark lines appear. If we introduce, for example, an alcoholic solution of the anilin colour 'Fuchsin' into a hollow prism (fig. 53) and look through it at a brightly illuminated slit, we obtain a spectrum in which blue and violet are less deflected than yellow and red. What is elsewhere the end of the spectrum here appears at the commencement, towards the middle it fades, and in the centre the green, being absorbed, is absent (fig. 140). From this behaviour the conclusion may be drawn that in Fuchsin the blue and violet rays are propagated with greater velocity than the red and yellow.

This phenomenon, which was discovered by Christiansen, and was shown by Kundt to be presented by a great number of absorbing substances, has been called *anomalous dispersion of light.*

104. The phenomena of anomalous dispersion renders us strongly disposed to the opinion that neither the refrangibility nor the length of undulation, but the number of vibrations, is to be regarded as the characteristic of a homogeneous ray of light. The number of undulations by which the impression of colour perceived by our eyes is conditioned *does not undergo any alteration in the passage of light from one medium into another.* In fact we observe no change of tint (Tonhöhe) when, for example, the yellow light of Sodium passes from air into water.

The length of the waves, however, does undergo a change. The wave length is, it is to be remembered, always obtained by dividing the velocity of propagation by the number of vibrations. As the latter remains unchanged, whilst the velocity of propagation in water is only three-fourths of the velocity in air, the wavelength in water can only amount to three-fourths of the wave-length in air. *The wave-length of a ray of light in any given substance is consequently obtained by dividing the wave-length in air by the index of refraction of the substance itself.*

105. We possess no means of changing the number of vibrations, that is to say, the colour of a homogeneous ray of light. But that such an alteration may and does occur under certain circumstances may now be demonstrated.

The sensation of a definite colour is conditioned by the number of waves of æther that penetrate into the eye in a second, just as the pitch of a musical note depends on the number of waves of sound which enter the ear in the same space of time. As long ago as 1841 Doppler called attention to the fact that the pitch of a musical note or the colour of an impression of light must be raised or lowered when the resounding or luminous body approximates or recedes from the observer. In the former case the organ of sense is struck in the course of a second by a greater, in the latter case by a less, number of waves than if the source of light or sound be stationary. As regards sound the truth of the principle of Doppler can easily be demonstrated by experiment; it is only necessary to allude to what may perhaps have been noticed by many. During the passage of a train through a station it may be observed that the whistle of a locomotive becomes higher in pitch as it approximates to, and lower in pitch as it recedes from the observer than when it is at rest. It is impossible, no doubt, to make a similar experiment in the case of light, because the greatest velocity we can attain is vanishingly small in comparison with its enormous speed. Nevertheless the possibility of its occurrence in the case of the waves of light cannot be doubted.

Let it be conceived that in free space a sphere of glowing Sodium vapour is moving with sufficient velocity *towards* our earth, its light would appear more green than that of a terrestrial Sodium flame, whilst if it were *receding* it would assume a reddish tint. And if this light fell upon a prism instead of our eye it would reach the prism in the former case with a greater and in the latter case with a smaller number of undulations than that of a Sodium flame at rest, and in correspondence with this would experience a stronger or weaker deflection. Hence it follows that if a spectroscope be directed towards the moving source of light the bright Sodium line would appear to have changed its position and to be advanced towards the more, or towards the less refrangible end of the spectrum, according as the source of light was approximated to, or made to recede from the observer.

Just as the bright Sodium line in this example undergoes a change of position, so also, when the fixed star moves in the direction of the visual line with sufficient velocity, do the dark lines in the spectrum of a fixed star become altered and no longer coincide with the bright lines of the elementary substances to the absorbing action of which they owe their origin. From the direction and amount of this dislocation both the direction and the velocity of the movement of the star can be deduced.

106. Huggins, on comparing the F line of the spectrum of Sirius with the blue-green line of the spectrum of a Geissler's tube filled with hydrogen, found the former as compared with the latter moved

towards the red, and to an extent that if the wave-length of the Hydrogen line F is 486·5 millionths of a millimeter, the wave-length of the line of Sirius corresponding to it was about 0·109 millionth of a millimeter greater. Hence it appears that at the time of observation Sirius was receding from the earth with a velocity which, compared with that of light (300,000) is as 0·109 is to 486·5. The velocity with which the two celestial bodies receded from each other was consequently

$$\frac{300{,}000 \cdot 0\cdot109}{486\cdot5} = 67 \text{ kilometers} = 41\tfrac{1}{2} \text{ miles.}$$

Since at the time of the observation the earth was moving away from Sirius at the rate of 19 kilometers (nearly 12 miles) in each second, there remains *a surplus 48 kilometers* (about 30 miles), *which represents the rate at which Sirius was receding from our solar system.*

Observations made with the telescope long ago taught astronomers that many fixed stars possess a proper motion of their own. But it is obvious that by means of the telescope only that portion of the motion can be recognised which takes place at right angles to the visual line. The spectroscope, on the other hand, indicates to us the movement which escapes telescopic observation, that namely which takes place *in* the visual line. It is thus apparent that by combining the results of different methods of investigation it is possible to determine the true motion of the fixed stars in space.

The conclusion may be drawn from the rapid change of form which is observable in the solar protuberances that the glowing masses of hydrogen of which they consist are in the most violent motion. Peculiar dislocations and disturbances, which Lockyer has observed in the dark F line of the solar spectrum as well as in the bright F line of the photosphere, have enabled him to measure with precision the velocity with which the glowing hydrogen streams up in the solar atmosphere or revolves in whorls of storm.

The alteration in the number of undulations which corresponds to these dislocations shows that a rapidity of from 50 to 60 kilometers per second is nothing unusual, indeed the most marked dislocation hitherto observed indicates a velocity of 190 kilometers (nearly 120 miles). If we compare these fearfully violent hydrogen storms in the solar atmosphere with even the most violent hurricanes of our atmosphere, which at most do not exceed 45 meters (=150 feet) per second, these last appear to be only gentle breathings.

107. We have hitherto regarded the process of the propagation of light from the standpoint of the undulatory theory. It remains to consider *the origin of light*, that is to say, the process of illumination, from the same point of view. The analogy of light to sound, which has so often afforded us useful hints, will also aid materially in this part of the subject.

A mass of matter becomes a source of heat and light in consequence of an extremely rapid vibrating movement of its smallest particles, which is propagated as a series of undulations into the surrounding æther, and is felt by our tactile nerves as heat, but by our optic nerves, if the undulations are sufficiently rapid, as light.

Certain facts which especially belong to the domain of Chemistry lead to the conclusion that the matter of which bodies are composed does not entirely fill the space it occupies, but consists of separate particles which, in a physical sense, are no further divisible, and are therefore termed ultimate particles or *atoms*. The interspaces of the atoms are filled with æther.

There are as many different kinds of atoms as there are elementary chemical bodies. Chemistry teaches us further that the atoms in any substance, even in an element, never occur singly, but are always united by the action of chemical affinity to form groups of two or more atoms. Such a group of atoms is called a *molecule*. Every molecule is built up by its atoms in a perfectly definite manner. The kind, number, and grouping of the atoms which compose a molecule determine the chemical qualities of the molecule, and consequently also of the substance, which consists of an indefinite number of such similar molecules.

And just as a chord gives a definite fundamental note besides its overtones, dependent on the length, thickness, tension, and consistence of the chord, so also the atoms within every molecule are capable of only a definite series of vibrations, the number of vibrations being determined by the structure of the molecule, that is to say, by its chemical properties. And just as it may be said that a chord or tuning fork is tuned to give a particular note, it may also be said that a Sodium molecule is tuned to the colour-tone D.

It may hence be imagined that the chemical nature of a body must betray itself by characteristic bright lines in the spectrum of its light.

Whilst the *chemical* properties of a body are determined by the *internal structure* of its molecules, its *physical* properties, especially its condition of aggregation (whether it be solid, fluid, or gaseous) depends upon the special mode in which its molecules are arranged amongst themselves.

In a *solid body* the molecules are held together, in determinate positions of equilibrium around which they can vibrate, by a powerful force, which is termed the force of cohesion (Zusammenhangkraft). These vibrations are independent of the peculiar quality of the molecules; they include also some vibrations of a less known character, but take place with all possible numbers of vibrations, and for all solid bodies in a similar manner at the same temperatures.

Solid bodies therefore, whatever may be their chemical nature, give alike a *continuous spectrum*, which at a lower temperature only contains the invisible ultra-red rays; as the temperature rises not only does the strength of the radiation increase, but a higher tone of colour is constantly being superadded to that previously present. At about 540° C. the red shows itself as far as B (dark red glow, dull or low red heat); at about 700° C. (bright or cherry-red heat) the spectrum extends to the farther side of F; and lastly, at white heat (1200° C.) it reaches to H. Glowing fluids, between the molecules of which the force of cohesion still acts, exhibit a continuous spectrum. These vibrations which the molecules of solid and fluid bodies exhibit under the influences of the force of cohesion, do not prevent the simultaneous occurrence of those vibrations within each molecule to which the molecule is attuned owing to its chemical composition. As a general rule* the latter are not visible, because

---

* According to Bahr and Bunsen the fixed oxides of Erbium and Didymium, when heated to glowing, exhibit a spectrum with bright lines which correspond to the dark striæ in their absorption spectra. (See § 75.)

the bright lines which correspond to them disappear upon the bright background of the continuous spectrum. The characteristic linear spectrum which discloses to us the chemical quality of a body is much better and more clearly seen when its molecules, freed from the chains of cohesion, enter into the gaseous condition.

108. Fig. 141 represents a tuning fork fixed into a little wooden box open at one end, and when made to vibrate it is heard to give a pure soft tone. A second tuning fork similarly supported on a box is placed beside it. If now the first be made to vibrate and be then immediately silenced by touching it with the finger, the second one, which was previously at rest, will be heard resounding with the same note. It has been set into vibration by the waves of air which proceeded from the first.

FIG. 141.

Tuning fork.

But if the second fork be put out of tune by attaching a little piece of wax to its arms, and the experiment be repeated, it remains perfectly silent. The resonance thus only occurs when the two forks are in unison with each other, that is, when the second possesses the same number of vibrations as the undulations of air proceeding from the first.

A similar phenomenon is familiarly known to all. If a person sings into an open piano with a loud voice the same note is gently returned in answer; those chords namely, which when struck by their hammers yield this note, are set in vibration by the sound, but the waves of sound excited by the singer pass over all the other chords without acting on them.

This vibration in unison which is called forth by tones of equal height, and is termed *resonance*, may be easily explained. Every wave of sound which reaches the tuning fork begins to set it in movement. If the impulses of the waves succeed to each other in the same time as the vibrations of which the tuning fork is capable, each arm of the fork when it is about to move forwards will receive an impulse forwards, and when it moves backwards an impulse backwards. The succeeding impulses thus act unopposed to strengthen the movement which was only feebly commenced by the first, and soon excite the fork to lively vibration. If, on the contrary, the number of vibrations of the waves differs from that of the fork, the later impulses very soon come to be in opposition to the slight trembling excited by the first, and neutralise their action. The tuning fork therefore remains at rest. To set the tuning fork in motion the unisonal waves must give up a part of the energy of their motion to it; they therefore proceed in a weakened condition on the other side of the fork. The waves not in unison, on the other hand, give off none of their energy to the tuning fork, but pass by it of their original strength.

If now a large number of tuning forks be imagined to be attached to a table, and a sound wave unisonal with them be excited at one end, it will reach the other in a very weakened condition, because its energy will have been in great measure absorbed by the tuning forks. A wave of another pitch will, on the contrary, traverse the layer of tuning forks almost unaltered, and will spread beyond them without noticeable loss.

DISPERSION OF LIGHT. ABSORPTION. 253

A Bunsen's flame in which float glowing particles of Sodium is comparable to such a layer of tuning forks, and it is now intelligible why the peculiar kind of light, D, which it emits, is weakened or altogether vanishes in traversing it, whilst it remains transparent for all other kinds of light.

The undulatory theory thus affords an explanation of *absorption*, inasmuch as it shows that every body must absorb exactly those kinds of luminous rays which it is itself capable of emitting.

109. Although a wave vanishes by absorption, the energy of its movement is by no means suppressed, but is transferred without loss to the absorbing body. For in accordance with the fundamental law of all natural phenomena, the *principle of the conservation of energy*, energy can as little be destroyed as created.

The motor energy which is transferred to the absorbing body may become manifest in this in two forms; a clock can obviously be set and kept in motion if the axis of the great wheel be turned. In this case the active energy of the hand is transferred into the *active energy* of the clockwork in motion. A watch may also be made to go by winding it up, that is to say, by coiling an elastic spring around the main wheel. The *active energy* of the hand is now transferred to the wound-up spring, and remains slumbering in it as *inactive energy*, or *energy of tension*, as long as the movement of the clockwork is checked. But as soon as the detent is loosed, however long a period may elapse, the spring gradually uncoils itself to its previously unstrained condition, and thus the whole energy which had been concealed in it in an inactive state again makes its appearance as the active energy of the clockwork in motion.

Let this simile be applied to the absorption of the æther waves. A portion of the active energy of the absorbed wave sets the molecules, and the atoms within the molecules, in motion, or renders the motion already present in them more lively. They become themselves by this means the centre of waves of æther, the *active energy* of which betrays itself to our senses as heat or light (glowing phosphorescence and fluorescence).

Another portion of the energy absorbed is employed in loosening or altogether dissolving the chains which bind the molecules together to form a substance, or the atoms together to form a molecule. When the molecules of the body, or the atoms within each molecule, are widely separated from each other or are completely dissociated, the body becomes extended, and passes from the solid into the fluid or gaseous condition; or lastly, it experiences, if the molecules split into their atoms, a chemical decomposition. In the former physical, as in the latter chemical action, a portion of the absorbed energy is consumed in overcoming the molecular forces (force of cohesion and of chemical affinity), just as the energy of the hand applied in winding up the watch is used to overcome the elastic force of the spring. The energy so applied, is, however, by no means lost, but remains stored up in the body or in its particles as *energy of tension* as long as the body remains in its condition of solution or division. It makes its appearance immediately again as active energy, in its original amount, if the body revert from its new into its old condition.

110. The various operations which the radiation

from the sun can produce on the surface of our earth may serve to illustrate these statements. Were the sun's rays completely reflected from the surface of the earth they could neither warm nor in any other way act upon it; their action is only rendered possible by the absorbing action of terrestrial objects.

The transparent air allows the sun's rays to traverse it almost undiminished in intensity, and is therefore to only a very slight extent directly warmed by it. On the other hand, the solid crust of the earth, which possesses considerable absorptive power, undergoes a great amount of heating; the air itself becomes gradually warmed from the soil; and since this heating takes place unequally at different parts of the earth's surface, attaining for example a higher degree in the equatorial than in the polar regions, the equilibrium of the atmosphere is disturbed, and seeks restoration by currents which we call winds. The movements of our atmosphere are thus primarily caused by the sun's rays; in the breeze which swells the sails of the ship, as in the hurricane which uproots trees, a part of the energy is made manifest which the sun sent down to the globe of the earth in the form of æther waves.

The evaporation which takes place from the surface of the sea under the influence of the solar rays causes the ascent of extraordinary quantities of aqueous vapour into the higher regions of the atmosphere, from whence, again condensed, they descend, in the form of water or of snow, and collected into streams and rivers, flow back to the sea. In performing this circuit the water gives off the whole of the energy which it originally received from the sun. The falling drops of rain, the ship-bearing river, the waterfall which turns the mill-wheel or drives the tunnel-borer through the granite of the Alps, owe their energy to the sun.

In the green leaves of plants the carbonic acid they have absorbed from the air undergoes decomposition by the absorbed solar rays, and the oxygen returns to the air in a gaseous form, whilst the carbon is applied to the construction of the solid parts of the plant. In the wood of the stem of a tree the whole energy of the solar rays which has been consumed in its formation in the course of years is found stored up in an inactive condition; it reappears with undiminished intensity as active energy in the form of light and heat when the wood, or rather the carbon contained in it, again reverts by the process of combustion to the condition of carbonic acid. The Carboniferous strata, which are composed of the altered remains of ancient plants, represent a highly economical mass of solar energy which, after a slumber lasting for ages, is again set free by the *process of combustion*, heating and illuminating our houses, striking the hammers and turning the spindles in our workshops, and driving our locomotives with the speed of the wind along their iron paths.

Amongst the animal creation some feed directly on vegetables, whilst others consume their plant-eating congeners. In both instances we recognise the vegetable world as the only spring of all animal life. In the animal organism the carbon consumed as food unites with the inspired oxygen, and is exhaled in the form of carbonic acid. The force condensed in the vegetable streams forth again in the animal body; that is to say, the energy of the solar rays which the plant required for the separation of the carbon is again set free in the animal body as heat and motion. The heat of the blood, the motion of our heart, the capacity for work in our arms, all represent the energy which originally streamed from the sun. Thus the sun, by means of the waves which it excites in the æther ocean of the universe, is the origin of all the heat, life, and motion on the surface of our earth.*

* There are no doubt a few terrestrial movements which are not occasioned by the radiation from the sun; such, for instance, as the ebb and flow of the tides, which are caused by the force of attraction of the moon and sun upon the waters of the sea. So also volcanic activity which has its origin in the interior of the earth. Lastly, there are stores of energy of tension which do not depend upon the sun, which are stored up in certain combustible minerals (in virgin sulphur, iron, &c.). Nevertheless, all these sources of force together are very insignificant in comparison with those which are supplied to us by the sun.

CHAPTER XIX.

DIFFRACTION OF LIGHT.

111. The last four Chapters having been occupied in rendering the facts stated in the earlier section of this work intelligible on the undulatory theory, we may now enter upon the consideration of new phenomena of light adapted to support the views already expressed, and to supply additional means of determining the essential nature of light.

If a beam of parallel solar rays be allowed to fall upon a narrow vertical slit, and the transmitted light be received upon a paper screen at some distance from it, there is seen on either side of the bright line which naturally results from the shape of the slit, a series of alternate dark and light striæ (fig. 142), which rapidly diminish in intensity as they are more distant from the central line, and are fringed with the same subdued colours that have already been seen in the interference lines of Fresnel.

Fig. 142.

Diffraction or inflection image of a narrow slit.

This experiment furnishes the practical proof that light spreads not simply in straight lines, but, as Huyghens' construction shows, laterally also. It is, in fact, simply the realisation of the case already mentioned (§ 97), that a wave in its passage through an opening, whilst it is propagated directly as a principal wave, also sends forth elementary waves into the space which is protected from the chief wave.

The white line in the middle is that part of the screen which receives the principal waves, that is to say, here all the elementary waves or elementary rays proceeding from the various points of the aperture are found in unison, and support each other in the most complete manner. The elementary waves uniting in a laterally situated point of the screen—called diffracted rays—are not capable of an equally favourable co-operation, since, proceeding from the various points of the aperture they travel over various paths to the screen, and become according to the difference of their path, i.e., according to the distance of the point of the screen observed from the middle stria, sometimes in partial accordance, sometimes in complete discordance, and thus are produced alternately the bright and dark striæ observed upon the screen. This phenomenon, because it originates by the interference of inflected rays, is termed a phenomenon of diffraction. When monochromatic light is used, the dark lines appear of a deep black colour, and are closer to each other, as well as to the central bright line, in proportion as the wave-lengths of the source of light employed are smaller. With white light, therefore, only the central stria appears white, whilst the lateral striæ appear, for the same reason and in the same order, coloured, like the interference striæ of Fresnel.

If the slit be gradually widened the lines will be seen to become progressively narrower, till they ultimately become so fine as to be no longer perceptible. In order therefore to perceive the laterally spreading elementary waves, very narrow slits alone can be used; with wide apertures they are undoubtedly present, but the phenomena of diffraction are then so extremely delicate that they escape observation.

112. The phenomena of diffraction may also be seen with the naked eye, if a distant object be looked at through a minute aperture. They may be still more advantageously observed by employing a telescope, at the objective end of which (A, fig. 143) a tube (B), lined with leather, is attached for the reception of the wooden ring, C. A sheet of tin is let into the latter, in which is a small opening, d. The diffraction figures which then come into view present various forms,

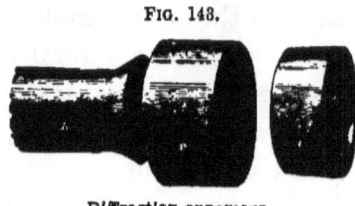
FIG. 143.
Diffraction apparatus.

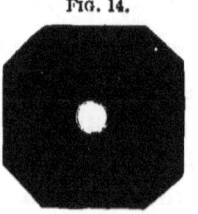
FIG. 144.
Phenomena of diffraction with a circular aperture.

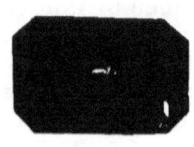
FIG. 145.
Phenomena of diffraction with a rhomboidal aperture.

according to the shape of the opening, and are often of surprising delicacy. Amongst the most simple is the figure which is obtained from a circular aperture (fig. 144). In this case a bright circular disc appears, surrounded by a succession of bright and dark rings, which, when white light is used, are fringed with delicate colours. With a rhomboidal figure (o, fig. 145) the image is divided by two rows of dark lines, each of which is parallel to the sides of the opening, into numerous parallelograms. The most distinct of these, which are arranged serially at the four sides, give to the image the aspect of an oblique cross artificially constructed in mosaic work.

When a telescope is used for the purpose of observing the diffraction image, it is formed in the focal plane of the objective, and is seen magnified through the ocular. The telescope permits consequently of the application of wider, and therefore of more strongly illuminated apertures, the diffraction figures of which would be too small to be seen by the naked eye.

113. It has already been pointed out how the phenomena of diffraction result from the interference of the elementary rays. It may now be advisable to enter a little more deeply into an explanation of them, under the supposition that they are being observed with a telescope, or even with the naked eye.

In fig. 146, AB represents the horizontal section of a screen, and C and D the edges of a vertical slit which has been made in it. If a fasciculus of parallel homogeneous rays, $c\,C\,d\,D$ fall vertically upon the screen, all æther particles within CD are in equal conditions of undulation. From each of them, in accordance with Huyghens' principle, elementary rays spread in all possible directions. All the rays which proceed from the various points of the aperture parallel to each other are united in one point of its focal plane by the objective. The fasciculus of diffracted rays, $CEDF$, for example, which forms the angle of diffraction ϕ with the axis CG of the incident rays, is united on a secondary axis parallel with CE, at the point where this

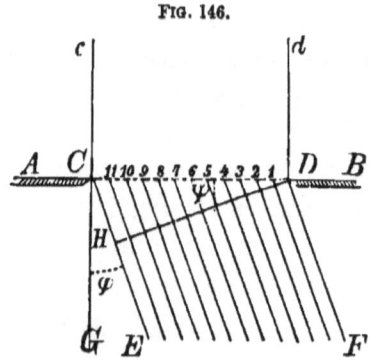

Fig. 146.

Explanation of diffraction taking place through a slit.

strikes the focal plane. The lens, however, as has been already pointed out (§ 102), exercises no influence on the difference of path of the rays within the fasciculus. These unite in the focal point with the same differences of path which were already present before it reached the lens.

If from the point D we let fall the perpendicular DH upon CE, CH constitutes the extent to which the path of the marginal ray CE exceeds the path of the marginal ray DF to the point of union. And for each of the ther innumerable rays of the diffracted fasciculus there is a portion between DC and DH, to which extent it falls behind the ray DF.

Those elementary rays which form the continuation of the incident rays do not indeed in any way differ from one another, and consequently meet in the chief focal point of the objective in the centre of the diffraction image in perfect unison. The larger, however, the diffraction angle becomes, and the more the diffracted fasciculus is inflected as regards the axis of the incident

DIFFRACTION OF LIGHT. 263

rays, the larger proportionately becomes the difference of path, CH, of its marginal rays.

With a certain small value of the angle ϕ, CH must be equal to half a wave-length of the incident homogeneous light. The marginal ray CE will then be in complete discordance with the marginal ray DF. These two rays must therefore neutralise each other at the point where they meet. The innumerable other rays of the fasciculus, on the other hand, have but little difference of path; they are not therefore in complete discordance with each other, but at the same time they are not in perfect accordance. A certain amount of light will therefore be present at their point of union, but this will be less than in the centre of the image.

If the angle of diffraction ϕ be so large that CH is equal to an entire wave-length, the middle ray (6) of the fasciculus is retarded a half wave-length as compared with the ray DF, and is neutralised by it where they meet. The same thing happens with the pairs of rays 1 and 7, 2 and 8, 5 and 11, which differ in their paths to the extent of a half wave-length. Since, consequently, every ray of the fasciculus finds a companion which is in complete discordance with it, darkness must prevail at the point where they meet. At this spot therefore, reckoning from the middle of the image, the first dark stria must occur.

If now, with still greater inclination of the diffracted rays, the difference of path of the marginal rays amounts to *three half* wave-lengths, it may be conceived that the beam is divided by the rays 4 and 8 into three equal parts. Thus the ray 8 is a whole wave-length behind the ray DF; the part of the fasciculus contained between them undergoes, as has already been shown, extinction, only the last third, the marginal rays of which differ by a semi-undulation, produces the effect of light at the point of union. But as this only contains a third of the whole amount of rays, whilst it otherwise exhibits the same difference of path as the entire fasciculus previously considered, with the marginal ray difference of a semi-undulation, the æther particles found at the point of union can only possess a three times smaller amplitude of vibration than the complete fasciculus. And since the intensity of light (see § 96) is always proportional to the square of the amplitude of vibration, it is obvious that the illumination at the point of union of the fasciculus having a difference of three half wavelengths in the marginal rays, is only the ninth part of that which the fasciculus with a difference of path of a half wave-length produces.

When with progressively increasing angle of diffraction the difference CH of the marginal rays is equal to *two entire* wave-lengths, the middle ray (6) remains a whole wave-length behind CF, and the ray DE a whole wave-length behind the middle ray.

Each half of the beam now has in itself the means of its extinction. Similarly, it may easily be comprehended that *every diffracted fasciculus of rays, the marginal rays of which differ in their path any number of whole wave-lengths, must disappear.* The dark lines in the diffraction image of the slit (fig. 142) correspond to these differences of path. The middle of the bright areas between each pair of dark striæ corresponds to the fasciculi whose marginal ray differences, 3, 5, 7 ... amount to an unequal number of half wavelengths. The intensity of light at these spots amounts to $\frac{1}{9}$, $\frac{1}{25}$, $\frac{1}{49}$ .... as compared with that which exists at those points where the difference of the marginal rays equals one half wave-length; these lie in the middle brightest area, which is twice the width of each lateral one.

114. In the right-angled triangle CDH (fig. 146) the angle at D is equal to the diffraction angle ϕ; if therefore the angle ϕ and the width CD of the slit be measured, we can easily estimate the length CH. The telescope of a Theodolite serves for the measurement of the angle ϕ (fig. 109). If it be first arranged in such a manner that its crossed threads are in the centre of the image, and it be rotated laterally till the first dark line appears at the crossed threads, the diffraction angle can be read off on the horizontal circle of the instrument; the corresponding value of CH must then be equal to the wave-lengths of the homogeneous light employed. Schwerd, for example, found that when red glass was used and the width of the slit was 1·353 mm., the first dark line corresponded to a diffraction angle of 1′ 38″, which gave for that particular red light a wavelength of 643 millionths of a millimeter.

Although the explanation we have given of the diffraction phenomena produced by a slit-shaped aperture refers only to the appearances presented when a telescope is employed, it will still hold for a diffraction image thrown upon a screen, if this be removed to such a distance from the aperture that all the rays passing to any point of the screen may be regarded as parallel to each other.

115. An inexhaustible variety of the most beautiful phenomena of diffraction may be produced by making a group of several or numerous apertures instead of a

single one. If, for example, a number of fine wires be stretched in a frame, the interspaces between them form so many slits, and we have a kind of grating. Such a grating of extraordinary delicacy may be obtained by cutting parallel lines at equal distances from each other on glass with a diamond. The lines drawn with the diamond correspond to the wires, and the unscratched surface of the glass to the interspaces of the wires.

If a fasciculus of solar rays be allowed to pass through the slit of a Heliostat and to fall upon a lens which projects a sharp image of the slit upon the adjacent screen, and if a fine glass grating be placed in front of the lens, a beautiful figure will become visible upon the screen (fig. 147). Symmetri-

Fig. 147.

Diffraction phenomena through a grating.

cally to the two sides of the white image of the slit a series of spectra appear, the violet end of which is turned inwards whilst the red is external. Whilst the two spectra on either side of the centre are isolated, the succeeding ones, which are progressively both broader and fainter, partially overlap each other. In these spectra, especially in the first and second on either side, the well-known lines of Fraunhofer are distinctly visible.

The same appearances are presented if the grating be held in front of the objective of a telescope

DIFFRACTION OF LIGHT. 267

placed at a little distance from the slit. On the supposition that this method of observation is adopted, an attempt may be made to explain the origin of these spectra.

In fig. 148 let AB represent the transverse section of the grating, and $M a N$ the direction of the incident rays falling vertically to the plane of the grating. All fasciculi of rays running parallel to each other, i.e., with the same diffraction angle ϕ at the bright interspaces of the grating, are united by the objective lens at the same spot of the image-plane. Disregarding for the moment the difference of path which exists amongst the elementary rays of each fasciculus, let us turn our attention to the difference of path of the several fasciculi in regard to each other. If from the point c, from which the first ray of the second fasciculus proceeds, a perpendicular, ch, be let fall upon the first ray of the first fasciculus, ah obviously represents the extent to which the first fasciculus is retarded as compared with the second, and consequently as each fasciculus is retarded as compared with the next succeeding one. If we now suppose the light to be homogeneous, as for example Sodium light, and the line ah equal to its wave-length, the whole of the fasciculi will be in complete accordance, and co-operate with one another at the point of union to give greater intensity of light. If the observer move to a very slight extent from that

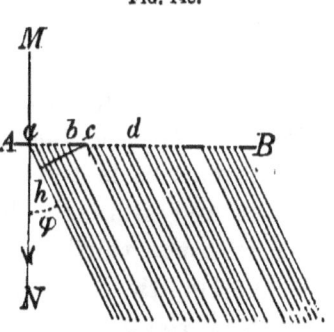

FIG. 148.

Explanation of diffraction through a grating.

position in which the difference of path of two adjoining fasciculi amounts to a whole wave-length, the fasciculi of rays must mutually extinguish each other at the point where they meet. If, for example, with a grating of 1,000 lines the angle of diffraction increases to such an extent only that the first fasciculus is retarded as compared with the second $1 + \frac{1}{1000}$ wave-lengths, it is retarded as compared with the third $2 + \frac{2}{1000}$; as compared with the fourth $3 + \frac{3}{1000}$, and so on until as compared with the 501st it is retarded to the extent of $500 + \frac{500}{1000}$, or $500 + \frac{1}{2}$ of a wave-length. The 501st fasciculus is thus in complete discordance with the first, and similarly the 502nd with the second, the 503rd with the third, and so on, until lastly the 500th with the 1000th. If, with a somewhat greater angle of diffraction, the difference of path between the first and second fasciculus were $1 + \frac{1}{100}$ wave-length, the 51st fasciculus would be in complete discordance with the first, and the fasciculi must again extinguish each other in pairs where they meet. Speaking generally, as soon as they recede on either side from the above direction, in which ah is equal to a whole wave-length, neutralisation of the waves occurs, providing only that the increase or decrease of ah is less than an entire wave-length.

For if ah be equal to *two entire* wave-lengths, all the fasciculi are again in complete accordance, and so on each occasion the difference of path of two adjoining fasciculi is equal to any number of entire wave-lengths.

The diffraction image perceived when Sodium light is used is consequently very simple. In the middle of the field of vision is the image of the slit; then follows at a certain distance on each side, which corresponds to the difference of path of a whole wave-length, a slender yellow line upon a perfectly black ground; then at double the distance, corresponding to the difference of path of two wave-lengths, is a second bright line, and others still at thrice, fourfold, &c. distances. One or the other of these pairs of lines can only then vanish when each of the fasciculi by which they are produced already carries in it the germ of neutralisation, that is, when the lines fall directly on the spots at which each interlinear space of the grating would exhibit a dark stria. Moreover, the diminution in the intensity of the light, which in general occurs in the consecutive lines, is to be ascribed to the interference which takes place in the interior of each separate fasciculus.

For every other homogeneous kind of light a series of bright lines of that particular light would be perceived, which, however, lie nearer the image of the slit when the wave-lengths are smaller, and on the other hand, more remote when the wave-lengths are greater. When white light is employed the striæ which correspond to the difference of path of each of the wave-lengths occur according to the serial succession of their wave-lengths, and form the first grating spectrum on each side of the white image of the slit; the second, third, and following spectra in the same way correspond to the greater differences of path. When certain kinds of rays are absent in the incident light it is obvious that hiatuses must exist at the corresponding points in the spectra, as for example at the Fraunhofer's lines when sunlight is used.

116. Owing to the occurrence of Fraunhofer's lines in the grating spectrum, we are in a position to deter-

mine accurately the wave-lengths corresponding to them. Fraunhofer himself, to whom we are indebted for the discovery of the grating spectrum, measured with the aid of the Theodolite the wave-lengths of the lines named after him with great precision. The *spectrometer* (fig. 110) is still better adapted for these measurements. If we place, for example, the grating instead of the prism upon the table of the spectrometer, and gradually focus the telescope upon the lines of Fraunhofer, the angle of diffraction corresponding to each focussing can be read off upon the divided circle. From the right-angled triangle $a\,c\,h$ (fig. 148), in which the angle at c is equal to the measured angle of diffraction, and the side $a\,c$ is likewise known as the sum of the breadth of a grating line and of an intervening space, the length $a\,h$ is obtained, which is equal to a wave-length, or is equal to two, three, and so forth wave-lengths, according as the measurement is taken in the first, or second, third, and so on, grating spectrum. The measurement of the spectra of the higher serial numbers serves to control the values furnished by the first spectrum. By means of this method Ångström has discovered the wave-lengths which are given in the following table in millionths of a millimeter:—

A	760,4	B	686,7	C	656,2
D_1	589,5	D_2	588,9	E	526,9
F	486,0	G	430,7	H_1	396,8
		H_2	393,3		

117. By means of the grating we have acquired a knowledge of the mode in which compound light may be broken up into its homogeneous components without any assistance from the refraction and dispersion of

DIFFRACTION OF LIGHT.

colour produced by a prism. The grating spectrum is therefore free from the influences which the nature of the material of which the prism is composed exercises upon the arrangement of the colours in the prismatic spectrum. In a grating spectrum the several homogeneous rays are arranged *essentially according to the differences of their wave-lengths* in air, and therefore according to a property which is inherent in the rays themselves.*

The grating spectrum is therefore to be regarded as the normal spectrum in which the position assignable to each homogeneous ray in consequence of its wave-length is not in any way altered by foreign influences.

FIG. 149.

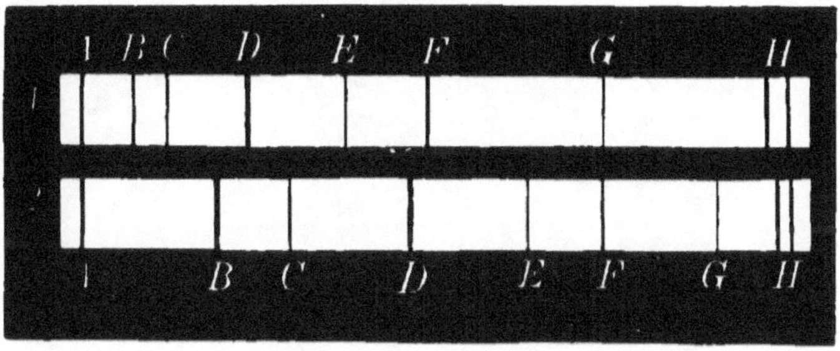

Comparison of the prismatic with the grating spectrum.

A comparison of the prismatic spectrum with a grating spectrum of equal length (fig. 149) enables the influence which the colour-dispersing material exercises upon the arrangement of the colours to be recognised.

* The number of undulations is always to be regarded as the distinguishing characteristic of a homogeneous colour. In the propagation of light in *free æther and in the air*, which occurs with equal rapidity for all kinds of rays, the number of undulations is always inversely proportional to the wave-lengths, and these may therefore be regarded as characteristic as those of the pitch of tone.

The middle of the grating spectrum is obviously occupied by those colours, the wave-lengths of which are intermediate between those of the extreme visible rays A and H_2. The wave-length 576,8, which is exactly intermediate between the greatest, 760,4, and the smallest, 393,3, corresponds to the yellow behind D. This colour therefore appears in the middle of the grating spectrum, whilst in the prismatic spectrum it is displaced towards the red end. Owing to prismatic dispersion the deeper tints of colour are approximated to each other, whilst the lighter tints on the contrary are more widely separated than in the colour scale of the grating spectrum at the same time rising progressively with the wave-lengths.

CHAPTER XX.

COLOURS OF THIN PLATES.

118. THE lovely play of colours on the soap bubble, well known to all from the happy days of childhood, long ago excited the attention of the physicist. Hooke more than 200 years ago was aware of the fact that every transparent body, if sufficiently thin, exhibited similar colours. He observed further that the fleeting colours of the soap bubble were arranged circularly around the thinnest spot of the fluid membrane, and he was soon successful in producing a permanent series of rings of colour by placing a very slightly curved plano-

FIG. 150.

Newton's colour-glass.

convex lens with its curved surface upon a plane glass plate (fig. 150). This simple apparatus, however, as well as the rings exhibited in it, are indissolubly associated with the celebrated name of Newton, because he measured the phenomenon and established the laws of the appearances presented.

If a large specimen of a Newton's colour-glass, showing the colours well, be taken, and a broad parallel beam of white light be allowed to fall upon it, whilst a lens is placed in the path of *the reflected rays*, a beautifully coloured system of alternately bright and dark rings (fig. 151) will be seen upon the screen behind the lens, which are more and more closely approximated from within outwards, and gradually become indistinct. The common centre of all the rings, is black. The colours from the centre to the first dark ring were named by Newton colours of the first order; from this to the second dark ring follow the colours of the second order, and so on. These colours are—

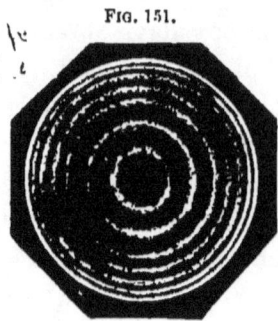

Fig. 151.

Newton's coloured rings.

First Order: black, pale blue, white, orange, yellow, red.
Second Order: violet, purple, yellowish-green, yellowish-red.
Third Order: purple, indigo, green, yellow, rose, carmine.
Fourth Order: bluish-green, yellowish-red, pale red.
Fifth Order: pale green, white, pale red.

If the lens be placed behind the colour-glass so that it now receives the transmitted rays, a system of rings is still seen upon the screen, the colours of which however are much fainter than they were previously in the reflected light. The centre of these rings is white, and their colours are arranged in complementary succession to those of the reflected rays. When homogeneous light is employed—if for example the incident rays be allowed to pass through a red glass—the rings appear in both cases merely alternately bright and dark; and in the transmitted light it may be observed that the dark rings occupy exactly the position of the bright rings in the reflected light.

COLOURS OF THIN PLATES. 275

119. An attempt will now be made to explain the mode of origin of these phenomena. In fig. 152, let $MNPR$ represent a thin layer of a transparent substance—for example, a piece of thin glass—upon which a beam of parallel rays falls in the direction ab. Every ray, ab, is in part reflected at the anterior surface, towards o, whilst it is in part refracted towards d; at d, before it leaves the lamina in the direction dh, it undergoes a second reflexion; and at the posterior surface, PR, a portion of the light here reflected reappears parallel with bo at the anterior surface.

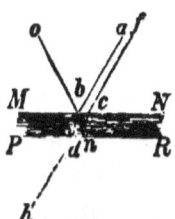

Fig. 152.

Explanation of the colours of the laminæ.

Disregarding the transmitted portion of each ray, dh, let us in the first place consider the rays which leave the plate parallel with bo after being reflected in part at the anterior surface MN, and in part at the posterior surface PR.

For each incident ray, ab, which is reflected at the anterior surface towards o, there is an adjacent ray, fc, the portion of which reflected from the posterior surface at n, on emerging from the anterior surface, follows the same path, bo. Of the two rays which pursue the same path, bo, the second, because it has had to traverse the path cnb within the film, is retarded, as compared with the other; to this retardation, which is obviously greater in proportion as the film is thicker, there is still a further retardation, dependent on the circumstance that the one ray is reflected in the denser, the other in the rarer medium; the reflexion in the denser medium, as has already been shown (§ 101), leading to a retardation of a half wave-length.

If, for example, the retardation within the film

amounts to a half wave-length of the red light, the two rays coursing along the line *b o* are in complete accordance, because in being reflected the one is retarded a half wave-length; the film therefore, if it be illuminated with red light, appears to an observer at *o* bright. The same would occur when the films are of three or five times greater thickness, because in these a retardation amounting to 3, 5, and so on, half wave-lengths occur. On the other hand, films which, on account of their thickness, bring about retardations equal to a number of whole wave-lengths, and which are consequently 2, 4, 6 . . . . times as thick as the first-considered film, appear dark with red light, because the two rays coursing towards *b o*, since they differ in their path by an unequal number of half wave-lengths, are in discordance. Were the incident light white, a film which retards red light a whole wave-length would extinguish the red, but not the other colours, the wave-lengths of which are different. The film would consequently exhibit a greenish tint, resulting from the mixture of all colours not extinguished; and were another film sufficiently thin to extinguish the yellow rays, it would appear blue with white light, and so on.

120. A film of perfectly equal thickness throughout will consequently exhibit the same colour in its whole extent—that, namely, which corresponds to its thickness.

In Newton's colour-glass we have to do with the film of air intervening between the two glasses, the thickness of which, proceeding from the point where the convex lens and the glass plate are in contact, increases in all directions from the centre outwards. At the point of contact itself, where the thickness of the film and consequently also the difference of path depending upon it, is *nil*, there is only a difference of path of a half wave-length, caused by the dissimilar reflexion of the two rays; there, consequently, is an extinction of light, and we see at this point a dark spot. If we pass outwards from the point of contact we meet with successive spots where the total difference of path for every homogeneous colour amounts to 2, 3, 4, 5 . . . . half wave-lengths, and where, consequently, alternate increase and extinction of light must occur. Thus we obtain an explanation of the system of rings with dark central point, even with homogeneous light. The smaller the wave-length the narrower must the rings be. When white light is used, neither the bright nor the dark rings of the different colours can therefore coincide, but in every concentric circle proceeding outwards from the centre, the colour resulting from the mixture of all the colours which have escaped extinction must make its appearance.

Let us now return to the thin lamina $MNPR$ (fig. 152), and consider the ray dh which leaves the film after it has traversed it along the line bd. With it is also associated a second ray, which after it has been reflected along the path $fcnbd$, and at n and b has been reflected inwards, has undergone a retardation compared with the others which corresponds to the length cnb. Since two reflexions take place either in the denser or the rarer medium, they either cause no difference of path or produce together a difference of a whole wave-length, and alter therefore in no degree the amount of coincidence or of opposition of the two rays which the film occasions in consequence of its thickness. The transmitted rays are consequently in complete accordance when the reflected rays are in discordance, and *vice versâ*. We see therefore in Newton's colour-glass, with transmitted homogeneous light, a bright centre and bright rings at those points where with reflected light the centre and the rings are dark; and in the same way with white light illumination the mixed colours are in the latter case complementary to those in the former.

But why is it that the rings appear so very much paler by transmitted as compared with reflected light? The answer is easily given; of the two rays which run in the direction dh, one, on account of its having undergone two reflexions, is much fainter than the other. The two rays therefore, even when they are in complete discord, can never entirely extinguish one another. On the other hand, the two rays reflected towards bo, of which each has been once reflected, are of nearly equal strength, and must consequently, as often as their difference of path amounts to an *odd* multiple of a half wave-length, undergo complete extinction. It is plain that the liveliness of the colours depends on the completeness of the interference.

121. But even in the reflected rings, as we proceed from within outwards, and as the film of air becomes progressively thicker, it may be observed that there is a decided diminution in the brilliancy of the colours, until the most external pale rings gradually disappear in a uniform white. It is easy again to explain why a thicker film exhibits no colours, appearing when illuminated by white light simply white. Let it be granted, for example, that a film is just so thick that it retards the red rays (B) about ten wave-lengths, apart from the retardation of a half wave-length which depends

on the dissimilar reflexions. Now since in the same length which includes 10 red waves there are about 17 wave-lengths of violet, the same film causes a difference of path in the violet rays amounting to 17 wave-lengths. Between the former red and these violet rays there are still other rays with 11, 12, 13, 14, 15, 16 wave-lengths in the same space which contains 10 waves of the B-red. The colours which correspond to these rays are in succession orange, greenish-yellow, green, bluish-green, bright-blue, indigo. All these rays must disappear in reflected light because the film causes in them a difference of path of an *odd* number of half wave-lengths. Those rays, however, the wave-lengths of which are contained $10\frac{1}{2}$, $11\frac{1}{2}$, $12\frac{1}{2}$, $13\frac{1}{2}$, $14\frac{1}{2}$, $15\frac{1}{2}$, and $16\frac{1}{2}$ times in the given length, because they strengthen each other, are seen of great brightness in reflected light. But to these the following colours correspond in succession: bright-red, yellow, yellowish-green, dark-green, blue, indigo-violet. An observer looking at the plate must obviously receive from the mixture of these colours the impression of white light.

122. That the colours first named are really absent in reflected light may be easily demonstrated by the following experiment. The solar rays are to be allowed to fall upon a plate of Mica, which to the naked eye appears white. The reflected rays are then made to traverse a slit, and are dispersed into a spectrum by means of a prism in the usual way. In this spectrum, between the red and the violet, *eight dark striæ* (fig. 153) are perceptible, corresponding to those kinds of rays which are extinguished by interference. A thicker plate of Mica is now to be selected, and the spectrum

now presents a very great number of dark interference lines (Müller's lines).

Fig. 153.

Interference striæ in the spectrum.

The spectrum of the light reflected from the Mica plate may be received upon a paper screen painted over with solution of quinine, and thus rendered fluorescent; and it will then be observed that in the now visible ultra-violet part of the spectrum such dark interference striæ make their appearance. And just as from the relation of the wave-lengths of red and violet the number of lines within the visible spectrum was formerly determined, we are now able, conversely, from the number of the lines that we can count from the violet to the end of the ultra-violet, to determine the ratio of the wave-lengths of the extreme ultra-violet rays to those of the violet rays, and consequently as the wave-lengths of all visible rays are known, to determine the wave-lengths of the most refrangible ultra-violet rays.

By an experiment essentially similar to the above, Esselbach found that the wave-lengths of the line R amount to 309 millionths of a millimeter.

Becquerel received the spectrum of solar light reflected from a film of Mica on a screen covered with a phosphorescent substance, and was able to follow the interference lines into the ultra-red region, where the

COLOURS OF THIN PLATES.

rays act in the peculiar manner mentioned above (§ 81), namely, apparently conversely to the more refrangible part of the spectrum. From the measurements obtained it resulted that the wave-lengths of the extreme ultra-red rays in this way rendered visible are more than twice the length of the extreme red rays. According to another less direct method, Müller determined the wave-lengths of the extreme ultra-red at about 4,800 millionths of a millimeter.

In music one note is said to be an octave above another if the number of its undulations is double, and consequently its wave-lengths half as great as the latter. If the same nomenclature be employed in the matter of colours, it may be said that the visible spectrum from A to H does not occupy a complete octave, but reaches from the fundamental note C to the sharp sixth $a\,i\,s$. If, however, the solar spectrum be considered in its whole extent, we find in the ultra-red alone, according to Müller, more than two octaves, to which must be added more than another octave from A to the line R in the ultra-violet. *The whole length of the solar spectrum thus embraces consequently about ~~four~~ octaves.*

CHAPTER XXI.

DOUBLE REFRACTION.

123. When after almost two thousand years of vain attempts on the part of the most accomplished mathematicians from Ptolemy to Kepler, to discover the law of refraction, i.e. the geometric relation between the incident and the refracted ray (see § 30), it was at last discovered in the year 1620 by Snellius, the ingenuity of observers was taxed afresh in 1669 by the ' wonderful and extraordinary refraction of light' discovered by Erasmus Bartholinus, Professor of Geometry in Copenhagen, in the beautiful crystal spar from Iceland.

The completely colourless and transparent calcareous spar depicted in the adjoining figure is bounded by six natural plane crystalline surfaces, of which the opposite pairs are parallel to each other. If a beam of parallel rays fall perpendicularly upon one of its surfaces, *two* such beams are seen emerging from the opposite one, which form upon a screen so placed as to intercept their passage two equally bright white spots.

This phenomenon is termed *double refraction*, and since in general every ray of light traversing the spar is split into two rays, all objects seen through such a crystal are doubled.

DOUBLE REFRACTION.

One of these two fasciculi, as it emerges, follows precisely the same course as the incident one would, if it traversed an ordinary plate of glass. The other, on the contrary, is laterally displaced in a direction which is dependent on the position of the crystal. If the crystal be rotated without altering its position in regard to the incident light, the bright spot which belongs to the first beam remains in its place, whilst the spot formed by the second beam, following the movement of rotation, describes a circle around the other.

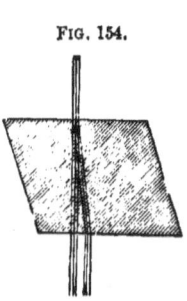

FIG. 154.

Double refraction in Iceland spar.

Again, if the crystal be gradually inclined more and more to the incident rays, the first beam exhibits nothing extraordinary in its behaviour, but constantly pursues the direction it ought to have in accordance with Snellius' law of refraction. These rays are consequently termed the *ordinarily* refracted ones. The other beam does not obey this law; it neither remains constant in the plane of incidence, nor is there an inalterable ratio between the angle of incidence and the angle of refraction; this ray is consequently said to be *extraordinarily refracted*.

124. The index of refraction of the ordinary rays may be determined in the usual way by means of a prism cut from the crystal. It is then found that its ratio of refraction (for Sodium light) is 1·6585. This number indicates that the velocity of propagation of light in air, as compared with the velocity of the ordinary ray in the crystal, is as 1·6585 to 1, or that if the former be equal to unity, the latter is 0·603.

The law of refraction of the extraordinary ray is

somewhat complicated. From the experiment above made, the conclusion may in the first place be drawn that the path of these rays stands in a certain relation to the form of the crystal. In order to investigate this law we must therefore first consider with some attention the crystalline form of calcareous spar. Fig. 155 represents the transparent model of a cube formed of twelve rods of equal length which are united at their extremities by hinges. If the cube be placed upon one of its angles, *a*, and the opposite angle, *b*, be pressed with the finger, the whole form of the model undergoes a change, the two compressed angles, *a* and *b*, become more obtuse, but the other six angles more acute than before, and the six originally square surfaces change into diamonds or rhombs; the cube thus altered is called a rhombohedron. Such a rhombohedron is the primary form of Iceland spar (fig. 156, *a*). The straight line, *a b* (fig. 155), connecting the two obtuse angles, is characterised by the circumstance that the surfaces, edges, and angles are arranged symmetrically around it. It is therefore called the axis of the crystal.

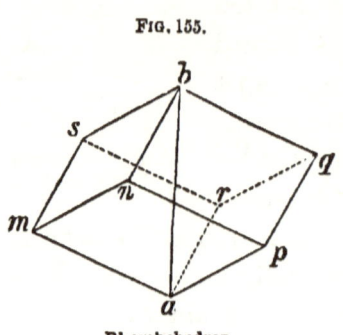

FIG. 155.

Rhombohedron.

The surfaces and borders are inclined equally to the axis, and the points of the angles and borders leading to them are equally distant from it.

Crystals of Iceland spar are not unfrequently found in which the six equal acute angles are replaced by six planes paralled to the axis of the crystal. The six-sided columnar prisms with rhombohedric ends, shown in fig. 156, *b*, originate in this way.

Crystalline forms of Iceland spar.

In other instances the obtuse angles have disappeared, and are replaced by surfaces which are perpendicular to the axis. We have then a six-sided columnar crystal with straight terminal planes (fig. 156, *c*). By cutting down the right and left edges, whilst leaving the anterior and posterior edges of the column as well as the two terminal surfaces, a rectangular parallelopiped is obtained, the upper and lower surfaces of which are at right angles to the crystalline axis, whilst the remaining four surfaces are parallel to it.

125. If now a thin beam of light be allowed to fall vertically upon the anterior and posterior surfaces of such a crystal, the axis of which is vertical, it will be seen that a *single* ray emerges from the opposite parallel surface in direct continuation of the incident beam. As soon, however, as the crystal is turned upon its axis, so that the beam strikes more and more obliquely upon its anterior surface, the double refraction becomes more and more obvious; and it may at the same time be remarked in regard to the bright spots upon the screen, that the two rays into which the beam is divided remain constantly in a plane perpendicular to the axis. Exact investigation shows further

that in this case, i.e., *when the plane of incidence is at right angles to the axis of the crystal, both rays follow Snellius' law of refraction.* If therefore a prism be cut from a piece of Iceland spar in the manner indicated in fig. 156, *d*, the refracting edge of which, *e f*, is parallel with the axis of the crystal, the ratios of refraction of both rays may be determined by means of it in the usual manner. For the more strongly deflected ray we find, as before, the number 1·6585, by which the ordinary refracted ray is characterised. The less refracted ray, on the other hand, which although in this particular case it is refracted in the ordinary manner, must be estimated as the extraordinary ray, gives the ratio of refraction 1·48654. It follows from this that the extraordinary ray is propagated in a plane at right angles to the axis of the crystal with a velocity of 1 : 1·48654, or 0·673, whilst the velocity of the ordinary ray is only 0·603.

As the two rays obey the ordinary laws of refraction, the construction can be applied to them from which we deduced (§ 98, fig. 138) the law of refraction itself. If, namely, two circles be described around the point of incidence, *a*, with the radii 0·603 and 0·673 (fig. 157), these will represent the contours of the *two* elementary waves contained in the plane at right angles to the axis, which have spread from the point *a* in the crystal in the time in which the light has traversed the length of path represented by unity in the air. If *a o* be any ordinarily refracted ray, and we draw to the point *o*, where it cuts the

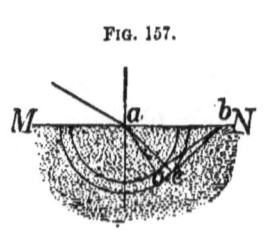

Fig. 157.

Double refraction. First case.

DOUBLE REFRACTION. 287

first circle, a tangent, $o\,b$, which strikes the surface of the crystal MN at the point b, we find the corresponding extraordinary ray when we join the point a with the point e, in which a straight line, $b\,c$, drawn from b, touches the second circle.

From this construction, the results of which agree in all points with observation, it follows however that the apparently simple beam which, when its incidence was normal, was seen to leave the crystal, really consists of two fasciculi which have traversed the crystal in the same direction, but with different velocities.

126. If now the cube of Iceland spar be so placed that its axis is at right angles to the incident rays, a single beam is seen to emerge from the opposite surface as a continuation of the incident rays; and if the crystal be now, as before, slowly rotated round the axis so that the incident rays constantly strike its anterior surface more and more obliquely, double refraction is observed to occur, both rays remaining always in the horizontal plane of incidence. Thus the ordinary ray, as might be expected, behaves itself exactly as in the former case, but the less refracted extraordinary ray now no longer follows Snellius' law of refraction. If we would construct it by a proceeding similar to the foregoing, we must, as Huyghens has shown, instead of the second circle draw an ellipse the half of the major axis of which, $a\,p$, is at right angles to the axis of the crystal, and is equal to 0·673, but the half of the minor axis of which, $a\,n$, is in the direction of the axis of the crystal and is equal to 0·603 (fig. 158).

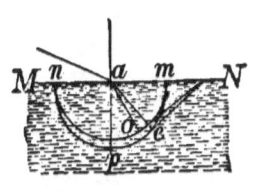

FIG. 158.

Double refraction. Second case.

20

In the plane of incidence parallel to the axis of the crystal the contour of the elementary waves corresponding to the extraordinarily refracted rays is represented by this ellipse, which touches the circular contour of the ordinary waves at the terminal points of its diameter which is parallel to the axis of the crystal.

The same ellipse in combination with the circle included in it also serves for the determination of the two refracted rays, when the incident rays strike at any given angle of incidence upon the terminal surfaces of the cube which are at right angles to the axis of the crystal, except that now the small axis, am, of the ellipse is at right angles to the surface of entrance, MN (fig. 159). When the light enters vertically it may also be observed in this third position of the crystal, as in the two first, that only a single ray leaves the crystal as continuation of the incident one; in the two first cases, however, this beam is only *apparently* simple, being in fact composed of two beams, propagating themselves in the same direction with different velocities; whilst in the case where it has traversed the cube in the direction of the crystalline axis, it is really simple, because in this direction the rapidity of propagation $am = 0.603$, is the same for the extraordinary as for the ordinary ray (fig. 159).

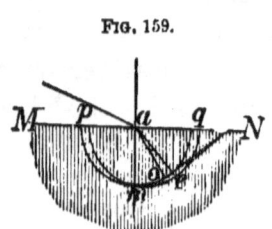

Fig. 159.

Double refraction. Third case.

Rays which pursue a course parallel to the axis of the crystal do not therefore undergo any refraction, whilst in every other direction two rays are propagated with different velocities. On account of this behaviour the axis of the crystal is also named the

DOUBLE REFRACTION. 289

optic axis. Every plane passing through the optic axis, or parallel with it, is termed a principal section. Thus, for example, the planes of the figs. 158 and 159 are principal planes, because they contain the principal axis within them. *All principal sections behave in exactly the same manner in reference to light.*

127. A view of the double-shelled elementary wave which spreads out from every point of a crystal of Iceland spar struck by light, in consequence of the two velocities of propagation, is obtained by combining the contours represented in figs. 157, 158, and 159 in an easily intelligible model (fig. 160). Since the ordinary rays are propagated in all directions with the equal velocity of 0·603, their wave-surface is obviously a sphere with a radius of 0·603. The wave-surface of the extraordinary rays exhibits, as we know, in every principal section, the same elliptical contour, ZXZ', ZYZ', the minor axis of which is coincident with the diameter ZZ' of the sphere that is parallel with the optic axis. It must therefore be represented as a spheroid flattened in the direction of the optic axis, but which everywhere includes the spherical waves of the ordinary rays, and only touches the optic axis in the terminal points Z and Z'. Whilst the axis OZ of the spheroid equals the radius of the sphere 0·603, the radius of its equator $(OX = OY = OX')$ amounts to 0·673.

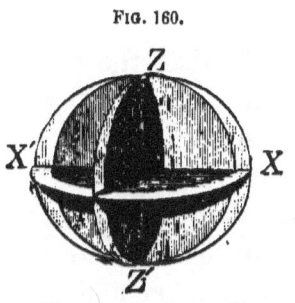

Fig. 160.

Wave-surface of a negative uniaxial crystal.

With the aid of these two-shelled wave-surfaces the two refracted rays corresponding to any incident ray may always be determined by a proceeding which is exactly similar to that applied to ordinary refraction in fig. 138. Fig. 161, which likewise makes it apparent upon the construction given by Huyghens for the case where the optic axis lies in the plane of incidence, but obliquely to the surface of the crystal, requires no further explanation.

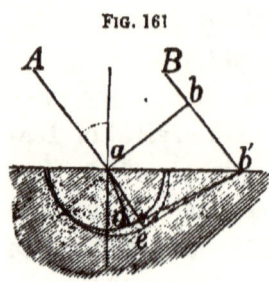

Huyghens' construction of double refraction.

128. The circumstance that the axis of symmetry of the crystalline form is also coincidently the axis of symmetry in relation to the propagation of rays of light, suggests that the cause of double refraction of Iceland spar is to be sought for in its special properties as a crystal.

Every crystal of Iceland spar is capable of cleavage parallel to the surfaces of its rhombohedric fundamental form (fig. 156, *a*) so that it may be easily broken up into smaller and still smaller fragments, the surfaces of which constantly maintain the same parallel position. These facts prove that the crystalline form is only the external expression of regular *internal structure,* which there can be no doubt is caused by a *certain orderly disposition of the molecules.*

All known crystals can be arranged in six great divisions or systems, in accordance with the laws which govern the grouping of their molecules. In the crystals of the *regular system*, the fundamental form of which is the cube, we find constantly three planes at right angles to each other (for example, the three edges that meet at any angle of the cube), which are completely similar to one another. Such crystals, like those of fluor spar and rock salt, exhibit no double refraction; they refract light in the same way as non-crystalline bodies, glass and fluids.

Two other systems of crystals, the *pyramidal* (das *quadratische*), and the *rhombohedral* (das *hexagonale*) possess one axis of symmetry developed beyond the others. All the crystals belonging to these systems are doubly refracting. Two rays are propagated in them in different directions, an ordinarily and an extraordinarily refracted ray. Double refraction is absent only in the direction of the axis itself, which is on this account named the *optic axis*. If the extraordinary rays have a greater velocity than the ordinary, the wave-shell corresponding to them has the form of a *flattened* spheroid which invests circularly the spherical wave of the ordinary rays. Crystals like Iceland spar, nitrate of soda, and others, in which this is the case, are termed *negative*. Those crystals are called *positive* in which, as in quartz, the ordinary rays possess the greatest velocity. In these the wave-shell of the extraordinary rays is represented by a *spheroid prolonged* in the direction of the *optic axis*, which is everywhere surrounded by the spherical ordinary wave, and is only in contact with it at the two extremities of the *optic axis*.

The crystals of the three remaining systems, the *right and oblique prismatic*, and the *anorthic* (rhombischen, klinorhombischen und klinorhomboidischen) are also doubly refracting, but neither of the two refracted rays—neither the retarded one nor the more swiftly propagated one,—obeys in general the ordinary law of refraction. In each of these crystals there are *two axes* without double refraction, which may be called the *optic* *axes.* These crystals are therefore termed *biaxially doubly refracting,* in order to distinguish them from the two preceding *uniaxially doubly refracting* systems. The wave-surfaces of the biaxial crystals consist also of two shells, of which one is enveloped by the other in such a manner that the two are connected with each other at four points corresponding to the terminal points of the two optic axes. With the aid of these wave-surfaces the direction of the two refracted rays can be determined in a similar way in biaxial crystals as in fig. 161 for uniaxial crystals.*

* [It must be observed that in this case the surfaces are not spheroids but surfaces of the fourth order.—TR.]

CHAPTER XXII.

POLARISATION OF LIGHT.

129. A BEAM of solar rays is constantly split into two beams *of equal brilliancy* by a crystal of Iceland spar, in whatever way this may be rotated round the axis of the incident rays. When Huyghens allowed these two rays to fall upon a second crystal of Iceland spar, he observed to his surprise that each was broken up into two rays of *unequal brilliancy*, the relative brightness of which depended on the position of the crystal, whilst there were two positions in which no double refraction occurred. From this phenomenon he rightly concluded that both of the rays refracted through a crystal of Iceland spar acquired peculiar properties, by which it was distinguishable from direct solar light. In order to distinguish conveniently one of the two refracted rays from the other, a natural rhombohedric crystal of Iceland spar may be employed (fig. 162, *A*), which is fastened by means of a cork ring in a metal tube. The tube is closed at both ends by a cover perforated at its centre by the round apertures *a* and

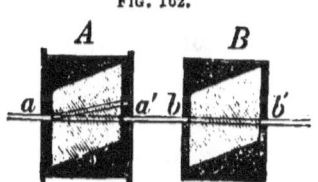

FIG. 162.

Two rhombohedra of Iceland spar.

a'. A second exactly similar rhombohedron of Iceland spar is attached to a tube (*B*), with a similar opening at both ends. If the tubes be arranged in such a manner behind each other that their axes are horizontal, and a beam of parallel solar rays be allowed to enter the aperture a in that direction, it will be seen that this, on account of its falling perpendicularly upon the anterior surface of the first crystal of Iceland spar, is split (as seen in fig. 154), into two rays, of which only the ordinarily refracted one emerges from the aperture a', and reaches the second crystal. In this position the principal planes of the two crystals running through the ray $a\,a'$ and the optic axis (the direction of which is indicated by the shading), lie in one and the same horizontal plane, namely, in the plane of the drawing.

In this *parallel* position of the principal planes, the ordinarily refracted ray emerging from the first crystal does not undergo double refraction afresh in the second crystal, but traverses it simply as an ordinarily refracted ray, without materially diminishing in brightness: as soon however as the second tube is rotated a little either to the right or left, double refraction takes place, and the extraordinarily as well as the ordinarily refracted spot of light appears upon the screen. As it is turned further and further, the extraordinary ray, which is at first only faint, continually gains in brightness, whilst the ordinary ray becomes proportionally fainter till both rays are of equal brightness, which occurs when the angle between the two principal planes of cleavage is 45°. On turning it still further the brightness of the ordinary ray progressively diminishes, and that of the extraordinary ray augments, till ultimately, when the principal planes of section are placed vertically, or at right angles to each other, the former has completely dis-

appeared, whilst the latter alone remains shining with
the full strength of the ray falling on the crystal B.
The ordinary ray again begins to appear as the rotation
is continued, and progressively gains in brilliancy with
the coincident and increasing faintness of the extraordinary ray until, after rotation to the extent of two right
angles, the principal sections of the two crystals again
coincide, when, as at first, the ordinary ray is alone
present in its original brilliancy. The same series of
phenomena are repeated when by further turning the
second crystal is rotated through two right angles, till
it arrives at the position which it originally had. The
ordinarily refracted ray emerging from the first crystal,
the principal section of which is horizontal, thus gives
rise either to an ordinary ray only, or to an extraordinary ray only, according to whether the principal
section of the second crystal is parallel, or at right
angles to it, and the double refraction which it undergoes in other positions takes place symmetrically on
both sides of the horizontal and of the vertical plane.

If every ray had the same properties around its axis
of movement, it would always produce the same phenomena whatever might be the direction in which the
second crystal of Iceland spar was turned. Its actual
behaviour however shows clearly that its upper and lower
sides are different from its right and left sides. Such
a ray possessing different sides is said to be polarised.

130. The knowledge of the fact that there are rays
of light with different sides, constitutes an important
step in advance in our enquiries into the nature of the
undulations of light. Hitherto we have only known
that the particles of æther arranged serially in the
direction of a ray of light performed to and fro move-

ments, but in regard to whether the direction of these vibrations takes place in the direction pursued by the ray itself, or forms an angle with it, the phenomena of light already considered afford no clue. However oblique to the direction of the ray the rectilinear vibration of an æther particle may be, we may, in accordance with the general laws of motion, regard it as composed of two vibrations, of which one, the longitudinal vibration, takes place in the direction of the ray, whilst the other, the transverse vibration, is at right angles to the ray. Consequently in regard to the direction of the vibrations in any ray of light, we have only the choice of three possibilities: they may be exclusively longitudinal vibrations, exclusively transverse vibrations, or coincidently longitudinal and transverse vibrations.

A ray of light which only presents *longitudinal vibrations* must exhibit everywhere the same characters around its line of propagation. This view therefore, since it is incapable of explaining the *laterality* of the polarised ray, must be unconditionally thrown aside. The phenomena of polarisation, on the other hand, can be at once explained if it be admitted that transverse vibrations are present. For if we suppose that in a horizontal ray of light, $a' b$ (fig. 162), the transverse vibrations only take place vertically upward and downward, but not sideways, its upper and lower side, towards which its vibrations are alternately directed, must obviously be different from its right and left side.

If now in the ray of light, $a' b$, longitudinal vibrations as well as transverse be present, they must pass through the second in the same way as they traverse the first crystal, whatever may be the position given to the latter.

But we have seen, however, that when the crystals are placed with their principal planes at right angles with each other the ordinary refracted ray disappears, and it may easily be demonstrated that at that spot of the screen where it ought to fall, not only is there no light but no heat, and no fluorescent action; the fact that at this spot where the longitudinal vibrations in case of their existence must necessarily fall, none of those actions occur which we now know to be characteristic of the æther waves, is most readily explicable on the assumption that in a polarised ray of light *there are no longitudinal, but only transverse vibrations.*

Fig. 163 represents consequently a *polarised* ray; the *plane* in which its transverse vibrations take place, the plane of the paper on which the figure is drawn, is called the *plane of vibration*. If a second plane be carried across the ray at right angles to the plane of vibration, the ray behaves itself symmetrically in reference to these two planes.

Experiment tells us that the refracted ray emerging from the first crystal of Iceland spar (the principal section of which is horizontal), is symmetrical in relation to planes carried through it in a horizontal and a vertical direction, but it does not tell us which of these two planes is the plane of vibration; and as other experiments directed to the solution of this question have not hitherto enabled us to give a decisive reply, we may accept whichever of the two planes we please as the plane of vibration. We prefer the vertical, that is to say, we admit that the vibrations of the ordinarily refracted ray are vertical or at right angles to the principal section of the crystal.

131. And now let the aperture a' be made in a small slide which can easily be placed in such a position that the extraordinary ray can alone emerge from the tube. If this be now examined in the same way as before by means of the second crystal, we see upon the screen when the principal planes are parallel the ordinary, and when they decussate at right angles the

Fig. 163.

Polarised ray of light.

extraordinary, ray. The extraordinary ray proceeding from the first crystal at once demonstrates itself to be polarised, and indeed *polarised at right angles* to the ordinary ray; that is, if we regard the vibrations of the ordinary ray as being at right angles to the principal plane, and *thus the vibrations of the extraordinary ray are in the plane of the principal plane of cleavage itself.*

132. The two polarised rays emerging from the Iceland spar contain, we must conclude, no longitudinal vibrations. The question arises whether the longitudinal vibrations are lost in their passage through the crystal from some absorptive action it possesses, or whether they are already absent in the direct rays of the sun. To obtain some data for an answer to this question, the first crystal, A, must alone be used, and the little cover a' must be removed from its frame. The two light spots belonging to the ordinarily and to the extraordinarily refracted rays then reappear upon the screen, and keep their original brilliancy in whatever

direction the crystal is turned. If the cover *a* be now removed, and its place supplied by another having a larger aperture, the light spots become correspondingly larger, though the space between their middle points is not changed; they are so large indeed that they partially overlap each other. In the part common to both, where, namely, the transverse vibrations of the ordinary mingle with those of the extraordinary rays, a degree of brightness is produced upon the screen which is not materially less than that of the direct light of the sun after passing through the same aperture without the intervention of the crystal of Iceland spar.*

If therefore longitudinal vibrations be present in the direct solar light, they nevertheless produce no obvious effect, or rather none of those effects which we have learnt to ascribe to the æther waves proper. The most probable view which presents itself in this respect is therefore that the unpolarised natural light, like the polarised, has no longitudinal vibrations, but consists *only of transverse vibrations*. This view receives essential support from the circumstance that all the known phenomena of light are only perfectly explicable on the assumption that light consists exclusively of transverse vibrations.†

* The slight diminution in the intensity of the light which may be demonstrated in the light which has traversed the Iceland spar, is fully accounted for by the two reflexions from the anterior and posterior surfaces of the crystal.

† It results from the laws of wave-movement that longitudinal vibrations, if present at all, must be propagated with unequal and greater velocity than the transverse vibrations, and consequently would already far outstrip them at even a small distance from the source of light. Since, moreover, the dispersion of colour can only be explained upon the admission of transverse vibrations, we are perfectly justified on these theoretical grounds in holding these last only to be luminous vibrations.

133. If the portion of light emerging from the Iceland spar common to the two beams be somewhat more closely examined, for example by allowing it to fall upon the second crystal, it will be found that it behaves just like natural non-polarised light. *By the combination of these two rays of equal brilliancy, polarised at right angles to each other, ordinary natural light is produced,* and conversely, every ray of natural light may be regarded as being composed of two equally bright rays polarised at right angles to each other. It is consequently of no importance what direction we assume for the vibrations of the one ray, if only it be admitted that those of the other equally bright ray are perpendicular to them. For the everywhere similar vibrations of the part common to the two beams present no variation in whatever manner the crystal is rotated around the axis of the draw tube; in every position the two-sidedness of the one ray is completely neutralised by the opposite two-sidedness of the other.

Two rays polarised at right angles to each other produce, as Fresnel and Arago have demonstrated by experiment, no phenomena of interference; they produce on the contrary (whatever may be their difference of path) always the same degree of illumination, that, namely, which is equal to the sum of the two rays in co-operation. It is evident in fact that two motions at right angles to one another cannot neutralise each other. Two polarised rays however, having a common source, that is to say, which originate in one and the same polarised ray, may clearly do so *if their planes of vibration coincide.* Two rays of natural light which proceed from the same source are therefore always capable of interference, for if either of them be conceived to be broken up into its two polarised constituents travelling in the two planes at right angles to each other, then those pairs of these four rays which have a common plane of vibration will act upon each other, and according to the amount of their common difference of path, coincidently abolish or strengthen each other.

134. The velocity with which a vibratory movement is propagated in an elastic medium is not simply dependent upon the density of the medium, but also upon the elasticity which this possesses in the direction of the vibration. In free space, in air, in water, in glass, and speaking generally in all simply refracting bodies, the elasticity of the æther is in all directions the same. The two constituents of a natural ray of light vibrating at right angles to each other propagate themselves therefore always with equal velocity, and remain throughout the whole of their path capable of being reunited to form a natural ray of light.

The mechanical disposition of the molecules in a doubly refracting crystal is the cause of its physical properties differing in different directions. In a crystal of this kind it may be demonstrated that heat is propagated with unequal velocity in different directions, that it expands unequally when heated, and that its various surfaces show different degrees of resistance to cleavage, and to the chemical action of various reagents. The view therefore appears to be justified that the elasticity of the æther contained between the molecules of the crystal is different in different directions. In crystals with an axis of symmetry for example, we must admit that the elasticity around and at right angles to the axis is of one kind, whilst it is different in the direction of the axis itself, and continually changes in passing from this direction into the former.

This theory renders it intelligible why the two components of a ray of natural light vibrating at right angles to each other in traversing a crystal of this kind, break up into two polarised rays which are propagated with unequal velocity. It is only when the ray of light follows the optic axis itself that *its two components* vibrate at right angles to this, and call into play equal elastic forces; they are hence propagated with equal velocity, and continue in their further path united to form a ray of natural light.

CHAPTER XXIII.

POLARISING APPARATUS.

135. In double refraction, which breaks up every beam of natural light into two polarised rays, we possess an excellent means of procuring polarised light. But inasmuch as the two beams when reunited are capable of again forming natural light, it is necessary to devise some method of setting one of them aside. This, for example, can be done by fixing a rhombohedric crystal of fluor spar, as in fig. 162, A, in a tube which is closed at its extremities with appropriate caps. In order that the ordinarily refracted ray may emerge separately from the tube, the diameter of each of the two openings, a and a', in the middle of the caps should amount to about only the tenth part of the thickness of the fluor spar. If these limits be overstepped, a portion of the extraordinary ray will also pass out through the opening a', and we shall no longer be dealing with completely polarised light. Applied in this way as a 'polariser,' even a very large crystal of Iceland spar can only give a very thin beam of polarised light. In order to employ this valuable material to greater advantage, Nicol conceived the following ingenious idea. He obtained, by cleavage from a crystal, a four-sided column with rhombic terminal surfaces, so that the form of the chief section through its obtuse lateral angles $a\,y$ and

e d, had the form of fig. 164. The prism is now to be sawn asunder along the line *b c*, that is, in the direction from one obtuse angle *e* to the other, at right angles to the principal cleavage plane, and the two cut surfaces, after they have been polished, are to be again cemented together in their original position by means of Canada balsam.

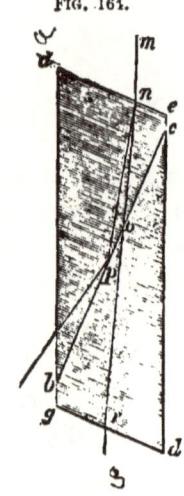

FIG. 164.

Nicol's prism.

If a ray of natural light, *m n*, fall on the rhombic anterior surface, *a e*, of the Nicol's prism,* it breaks up into an ordinary refracted ray, *n p*, and an extraordinarily refracted one, *n o*. The former, the index of refraction of which (1·6585), is greater than that of Canada balsam (1·53), strikes so obliquely upon the surface of the cement that it cannot penetrate it, but undergoes complete reflexion. The extraordinary ray, on the other hand, which propagates itself with greater rapidity in Iceland spar than in Canada balsam, penetrates the latter under all circumstances, and leaves the posterior surface, *d g*, as a completely polarised ray, *r s*, the vibrations of which, in my opinion, are parallel to the principal section, *a e d g*.

A Nicol's prism thus permits only those vibrations to traverse it that are parallel to its principal plane of cleavage, whilst it *is completely opaque for rays which are at right angles to the principal plane.* For the sake of convenience it is fixed in a metal frame, which is not usually provided with a diaphragm, for all the rays

---

* Often termed for the sake of brevity the 'Nicol.'

that fall parallel upon the first surface, mn, issue from the second as a completely polarised fasciculus of parallel rays, the breadth of which amounts to about a third of the length, ag, of the piece of spar employed.

The Nicol's prism, as well as, speaking generally, every 'polariser,' can also, conversely, be made use of as a 'polariscope,' that is to say, may serve as a means of recognising any ray of light as being polarised, and determine the position of its plane of vibration. If for example *natural* light fall upon a Nicol's prism, a polarised beam issues from it which maintains constantly the same degree of brilliancy in whatever manner the Nicol's prism be rotated around the direction of the incident rays. In fact, in every position of the Nicol's prism, half the incident light traverses it as polarised light. If on the other hand *polarised* rays be allowed to fall upon a Nicol's prism, they are only perfectly transmitted when its chief section is parallel with the plane of vibration of the incident rays; but if the Nicol be rotated out of this position, the transmitted light becomes constantly more and more faint, and ultimately entirely vanishes when the chief section of the Nicol is at right angles to the plane of vibration.

136. The Nicol's prism may now be applied as a polariscope to the investigation of the light reflected from a plate of mirror-glass which has not been silvered. A beam of natural light, ab, is allowed to fall upon the glass plate RS (fig. 165) at any angle, and is reflected towards c. If the Nicol's prism be placed in the path of the ray bc, and rotated around this ray as an axis, it will be observed that the transmitted light is sometimes brighter, sometimes fainter, though

it does not entirely vanish in any position of the Nicol's prism. The light reflected from the glass plate is consequently neither natural light, nor is it completely polarised. Its behaviour is rather as if it were a mixture of natural and polarised light, and it is therefore said to be *partially polarised*. The Nicol, in whatever position it may be placed, allows one half of the unpolarised constituent to pass through, whilst the polarised constituent is extinguished or transmitted according to whether the principal plane of the Nicol is at right angles to, or parallel with its vibrations. In order to determine the plane of vibration of the polarised portion, it is only necessary to place the Nicol in such a position that the transmitted light is as faint and feeble as possible. This takes place when the principal cleavage plane of the Nicol comes to lie in the plane of incidence, abc. From which we draw the conclusion, that the plane of vibration, $dflm$, of the polarised light contained in the reflected beam, *is at right angles to the plane of incidence, abc*.

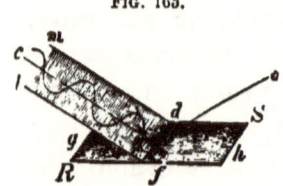

FIG. 165.

Polarisation by reflexion.

The proportion of the polarised portion to the non-polarised varies with the angle of incidence. With vertical incidence for example, the reflected beam contains no *polarised* light, but if the angle of incidence amount to 57°, or if the incident rays form an angle (abh) of 33° with the glass plate, the *unpolarised* portion is entirely absent. At this angle of incidence, which is known as the *polarisation angle*, the light reflected from the glass plate undergoes *complete polarisation*, and its vibrations take place at right angles to the plane of incidence (or parallel, df) as is indicated by the wave-line in the figure.

137. A glass plate placed at this angle, since it only reflects vibrations at right angles to the plane of incidence, thus forms an excellent 'polariser.' Instead of examining the rays reflected from it by means of a Nicol's prism they may be received at the same angle on a second glass plate (fig. 166), which then plays the part of a polariscope. If the two plates, as in the figure, are parallel to each other, their planes of incidence are parallel, and the ray bc, the vibrations of which are at right angles to the plane of incidence common to both, is reflected from the second plate to cd. But if the second plate be rotated from this parallel position whilst it still forms the angle 33° with the direction of the ray bc, the light reflected from it becomes weaker and weaker till it entirely disappears when the two planes of incidence are at right angles to each other. For in this crossed position the vibrations of the ray bc lie in the plane of incidence of the second plate, and are not reflected, because only those vibrations that are at right angles to their plane of incidence are capable of reflexion. In order to arrange this experiment conveniently, the apparatus shown in fig. 167 may be employed. To

FIG. 166.

Two polarising mirrors.

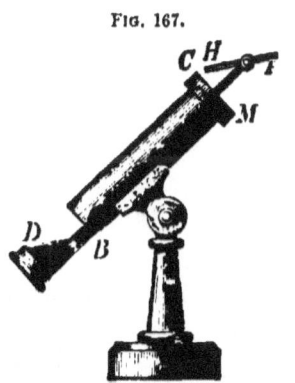

FIG. 167.

Biot's polarising apparatus.

one end of a tube blackened in its inside a mirror of black glass, DB, is so attached that it forms an angle of 33° with the axis of the tube. Rays which run parallel to the axis of the tube from D to C, are reflected at the mirror under the angle of polarisation, and are therefore completely polarised. A second blackened mirror is attached to a ring at the other end of the tube, which is likewise inclined at an angle of 33° to the axis of the tube, and by rotation of the ring can be brought into the different positions required in this experiment. A blackened mirror is selected in order to avoid transmitted unpolarised light, which might be mingled with the light polarised by reflexion. Silvered mirrors cannot be employed as polarisers, because they do not completely polarise the reflected light under any angle of incidence.

Every kind of apparatus which, like that just described, constructed by Biot, is composed of two polarising arrangements, of which one acts as polariser and the other as polariscope, is called a polarising apparatus. The apparatus of Nörremberg, shown in fig. 168, is the best adapted for the greater number of experiments. A transparent plate of mirror-glass, CD, here acts as a polariser, and forms, with the vertical axis, nc, of the instrument an angle of 33°; the light incident in the direction mn, which is completely polarised, is in the first instance deflected vertically downwards, and from thence it is reflected vertically upwards again upon itself by a mirror, c, fixed in the foot of the instrument, so that after it has traversed the glass plate, CD, it can reach in the direction of the axis of the apparatus the black mirror, $C'D'$, acting as polariscope. The ring i, to which two columns, a' and b', are

POLARISING APPARATUS. 309

attached, supporting this mirror, revolves within a fixed ring K, divided into degrees, and supported by the rods a and b. The zero of the divisions of the fixed ring is so arranged that when the indicator i of the rotating ring is placed upon it, the plane of incidence of the mirror $C'D'$ is parallel with that of the glass plate CD. In the present position of the instrument, the planes of incidence are at right angles (the indicator standing at 90°); the light coming from below is therefore *not* reflected by the mirror $C'D'$.

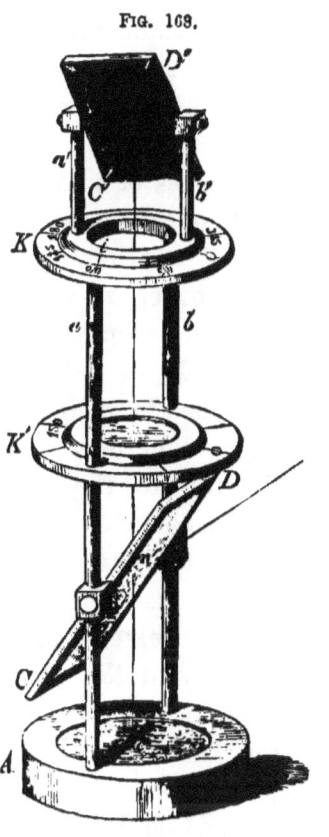

Fig. 169.

Nörremberg's polarising apparatus.

138. Moreover the light transmitted by a glass plate at an acute angle, when examined with a Nicol's prism is found to be *partially polarised*, and the vibrations of the polarised portion are constantly *in the plane of incidence*, or in other words the transmitted is polarised at right angles to the reflected light. As Arago has shown, the quantities of light polarised at right angles to each other in the refracted and in reflected rays are equal to each other at every angle of incidence. But whilst the reflected light at a determinate angle of incidence, namely, at the polarising angle, appears to be completely polarised, some un-

polarised is mingled with the transmitted light; it is always only partially polarised, whatever may be the angle selected.

In the same manner a nearly complete polarisation of the transmitted rays may be effected if, instead of a few, a sufficient number of glass plates be superimposed upon each other. If a ray of natural light fall upon such a series of plates placed at the polarising angle, and we conceive the same to be broken up into its two halves, of which one *vibrates in* the plane of incidence and the other *at right angles* to it, the former half, because on account of the direction of the vibration it is incapable of reflexion, is transmitted through all the laminæ almost without loss. The other half, on the contrary, undergoes at each surface a partial reflexion, and owing to these repeated reflexions becomes so faint as to be no longer perceptible. Of those rays which are presented to a succession of glass plates of this kind at the polarising angle, only such are transmitted, to any marked extent, as vibrate parallel to the plane of incidence, and the plates can therefore be used for a polariser as well as for a polariscope.

Fig. 169 shows a Nörremberg's polarising apparatus, the polariscope of which is the glass plate, CD. The light polarised by the glass plate AB, is extinguished when the plane of incidence is coincident with that of the series of plates, CD. This arrangement offers this advantage, that the visual line of the observer, whilst the polariscope is rotated, can remain constantly in the direction of the axis of the instrument, whereas in the instrument represented in fig. 168, the eye is compelled to follow the movements of the blackened mirror. The same object can also be more conveniently attained when the generally somewhat expensive Nicol's prism is applied as a polariscope.

'139. After Malus, in 1810, had discovered the polarisation of light reflected and refracted through glass plates, he showed further that almost all reflecting surfaces, with the exception of metallic ones, were capable of polarising light, but that the polarising angle at which this took place differed for different substances. That, for example, required in the case of Water is 53°; for Carbon bisulphide 59°; for Flint-glass 60°. From these values it appears that the polarising angle of any substance increases with its refracting power for light. Malus was, however, not in a position to ascertain this relation, and its discovery was reserved for the ingenuity of Brewster, who, in 1815, found, *that the polarising angle is that angle of incidence at which the reflected, forms a right angle with the refracted ray.*

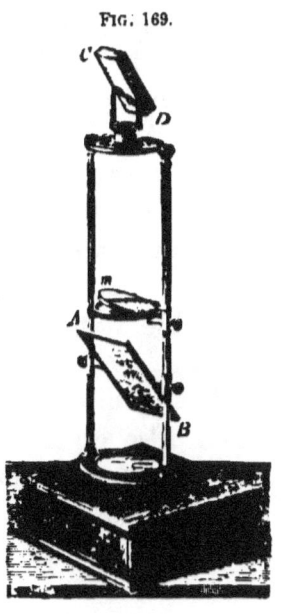

Nörremberg's polarising apparatus with glass laminæ.

This law supplies an additional means for the determination of the index of refraction, the more valuable since it can be used in the case of substances having only a small degree of transparency, and to which the former or prismatic method (§ 35) is not applicable. For just as by means of Brewster's law, we can deduce the polarising angle from the known ratio of refraction, so, conversely, we can obtain the ratio of refraction from the polarising angle.

The indices of refraction of Anthracite, 1·701; Horn, 1·565; and Menilite, 1·482, given in the tables, have thus been ascertained from observing the polarising angle. As the indices of refraction of the different coloured rays are unequal, their polarising angle, though perhaps only to a small extent, must also differ; white light can therefore never be completely polarised by reflexion, but only one of its homogeneous colours, whilst the rest only approximate to complete polarisation.

The undulatory theory, as Fresnel and Cauchy have shown, also gives an intelligible and satisfactory explanation of the phenomena of polarisation by reflexion and refraction. From the law of conservation of energy, which requires that the energy of the reflected and that of the refracted wave should be together equal to that of the incident wave, as well as from the condition that the amount of motion at the line of junction of the two media must be equal, we are enabled to calculate the nature of the reflected and of the refracted rays. From such a calculation the laws of Arago (§ 138) and of Brewster (§ 139), obtained by experiment, follow directly, and in all other respects it proves to be in complete unison with the results of observation.

140. As has been demonstrated, the colours of transparent bodies originate in the absorption which certain homogeneous colours, that is to say, rays of a definite number of vibrations, undergo in their passage through those bodies. In the case of coloured doubly refracting crystals the amount of absorption is dependent not simply on the number of vibrations of the transmitted

rays, but also upon the angle which the *direction of their* vibrations forms with the optic axis of the crystal, a circumstance which gives rise to a remarkable phenomenon which may now be investigated.

Let a small cube of Pennine, a mineral belonging to the rhombohedral system of crystals, in which the planes of two opposite surfaces are at right angles to the optic axis, whilst the others are parallel to it, be selected. If the observer look through the cube in the direction of the optic axis, it appears to be of a dark bluish green colour, whilst when looked at from the sides it has a brown colour. This peculiarity is called *dichroism*. These two colours will be seen on the screen if the sun's rays be transmitted through the crystal first in one direction and then in the other. The bluish green light which has traversed the crystal along its optic axis contains only those natural rays the vibrations of which are at right angles to the optic axis. The olive-green light, on the other hand, is composed of ordinarily refracted rays, which vibrate at right angles, and of extraordinarily refracted rays, which vibrate parallel to the axis. These two constituents may easily be separated from one another by a Nicol's prism placed behind the crystalline cube. For if the principal section of the Nicol's prism be placed at right angles to the optic axis of the cube of Pennine, the *same bluish green colour* appears upon the screen which was previously observed in the rays that had traversed the crystal in the direction of the axis, but if the Nicol be placed parallel to the optic axis, the bright spot upon the screen appears *brownish yellow*. The rays of light traversing a crystal of Pennine consequently experience an amount and kind of absorption varying according to whether their vibrations are at right angles to or parallel with the axis; in the former case they appear bluish green, in the second brownish yellow, and the above-mentioned brown is only the mixture of these two colours.

A remarkable inequality in the power of absorption according to the direction of the vibrations is shown by *Tourmaline*, which even when only of moderate thickness completely extinguishes ordinary rays.

A plate of Tourmaline, cut parallel to the optic axis of the crystal, allows therefore only the extraordinary rays vibrating parallel to the axis of the crystal to pass through it, and can therefore act as a polariser as well as a polariscope.

A combination of two Tourmaline plates, as shown in fig. 170, forming the so-called Tourmaline forceps or tongs, constitutes the simplest of all polarising apparatus. In this, for the sake of convenience, the plates are fastened by means of cork discs in wire rings, in which they can be made to rotate. By means of a coiled elastic wire they can be gently pressed together so that any object placed between them which is required to be seen with polarised light is held as if by a pair of tongs or forceps.

Fig. 170.

Tourmaline tongs.

If the plates be placed in such a position that their axes are parallel (fig. 171), the light of the sun traverses them just as it would through a single plate of the same thickness as the two together. But if one of the plates be rotated, the transmitted light becomes fainter and fainter, till when the axes of the two are at right angles it entirely disappears.

POLARISING APPARATUS. 315

The yellowish brown or brownish green colour which the Tourmaline communicates to transmitted light seriously interferes with its applicability as a

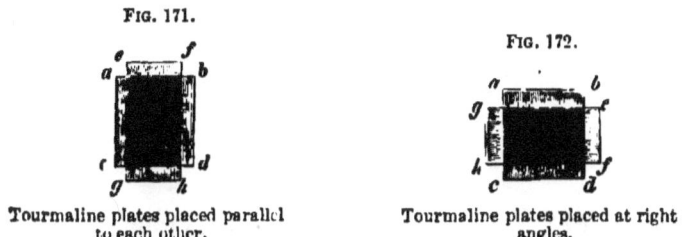

Fig. 171.
Tourmaline plates placed parallel to each other.

Fig. 172.
Tourmaline plates placed at right angles.

polarising apparatus, for which its simplicity would otherwise render it very well adapted.

CHAPTER XXIV.

INTERFERENCE OWING TO DOUBLE REFRACTION.

141. VERY few crystals exhibit the phenomena of double refraction so distinctly as Iceland spar; in most instances there is so small a difference between the two velocities of propagation that the splitting or decomposition of an incident beam into two fasciculi of rays can only be perceived when, as seldom happens, the crystals can be obtained of considerable thickness. The circumstance, however, that the two rays resulting from double refraction are *always polarised*, renders it possible to recognise even the slightest amount of double refraction, and to investigate its laws.

With this object in view, two Nicol's prisms, A and B (fig. 173), placed horizontally one behind the other, are employed as a polarising apparatus. The first, the principal cleavage plane of which is vertical, gives a parallel beam of vertically vibrating polarised rays which are not transmitted by the second, the principal cleavage plane of which is horizontal. The screen therefore is perfectly dark, the darkness continuing when a plate of any simply refracting substance, as for example glass

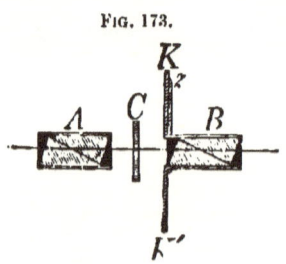

FIG. 173.

Two Nicol's prisms employed as a polarising apparatus.

INTERFERENCE OWING TO DOUBLE REFRACTION. 317

or rock salt, is introduced between the two Nicols. If, on the other hand, a lamina of a doubly refracting crystal, as for example a natural rhombohedron obtained by cleavage of Iceland spar, be placed at *C*, the screen appears *alternately dark and light* as the lamina is rotated around the axis of the rays.

This behaviour admits of an easy explanation. If a vertical line, MN (fig. 174), be conceived to be drawn upon the screen, the position of the principal cleavage plane of the first Nicol's prism, which serves as a polariser, is obtained; and in the same way the horizontal line, PQ, represents the principal cleavage plane of the second Nicol, which plays the part of a polariscope. The plate of spar is now introduced between the polariser and the polariscope, in the first instance in such a way that its principal cleavage plane coincides with the direction of the vibration, PQ, of the second Nicol. The rays emerging from the first Nicol, which vibrate parallel to MN, undergo only ordinary refraction in the crystalline plate. They traverse it without changing the direction of their vibration, and are extinguished by the second Nicol. In the same way extinction must also occur when the principal cleavage plane of the plate coincides with the plane of vibration, MN, of the first Nicol, for in that case all the rays pass as extraordinarily refracted rays through the crystal, whilst they preserve the original direction of vibration, MN. If the principal plane of the crystalline plate be brought into the position RS, it only allows, in accordance with the

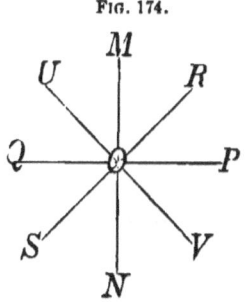

Fig. 174.

Decomposition of vibrations.

laws of double refraction, those vibrations to traverse it which run in RS, or at right angles to it, UV. The undulation, MN, as it emerges from the first Nicol, can now, since it forms an acute angle with the principal plane, RS, neither be continued completely in the ordinary nor in the extraordinary ray; on the contrary, it breaks up, in accordance with the laws of motion, into two undulations, of which one, running in RS, traverses the crystal as an extraordinary ray, whilst the other, vibrating at right angles to the principal plane (in UV), becomes an ordinarily refracted ray.

Two rays thus reach the second Nicol, of which one vibrates in RS, the other in UV. As the Nicol only transmits undulations which occur in its principal plane, PQ, each of these two rays is again divided into two parts, of which one vibrates in PQ, the other in MN. The two sub-rays whose undulations are at right angles to PQ are not transmitted by the Nicol; the two other sub-rays, however, which take place in its chief plane, PQ, penetrate it, and illuminate the screen.

We thus see that a doubly refracting plate, placed between two Nicols at right angles to each other, causes the field of vision or the screen to be dark in two positions, when its principal plane coincides with that of either of the two Nicols. In every other position light passes through it, and the screen is illuminated.* This behaviour is a positive proof of its doubly refracting nature.

142. Of the two sub-rays which, vibrating in the same plane, PQ, leave the second Nicol, the first

---

* Except only when the plate is cut at right angles to its optic axis.

originates the ordinary, the other the extraordinary ray, of which each propagates itself with its own velocity through the crystal plate. The one consequently lags behind the other to the extent proportional to the thickness of the lamina. In consequence of this difference of path induced by double refraction, the two rays polarised in a common plane of vibration occasion interference, which betrays itself when the difference of path is not too great, by beautiful colour phenomena.

The plate of Iceland spar used in the above experiment is too thick to show the effects of interference. If it be intended for this purpose, it must be rendered thinner by grinding. Crystallised gypsum, a biaxial doubly refracting crystal which cleaves easily into thin laminæ (Selenite) is a convenient substitute for the Iceland spar in these experiments on the phenomena of interference. If such a plate of Selenite be placed between the crossed Nicols, it behaves like the plate of Iceland spar; in two positions of the lamina, in a direction in which what we shall term its principal plane is parallel or at right angles to the direction of vibration (MN, fig. 174) of the polariser, the screen remains dark, but in every other position it exhibits *colours*, which are brightest when the principal plane of the lamina makes an angle of 45° with the axis of vibration of the first Nicol.

The lamina which is now in this position between the Nicols exhibits a beautiful *red colour*, originating in the interference of the two sub rays vibrating in PQ.

If the second Nicol be now rotated from the crossed position, the screen indeed continues to be illuminated, but the coloration diminishes in brightness, and is ultimately replaced by perfect white light, when the axis of vibration of the Nicols forms an angle of 45° with each other. If it be turned still further, a greenish colour appears, which finally, when the principal planes of the Nicols are parallel, becomes of a *bright green*. This colour is the result of the interaction of the two part-rays vibrating to MN. These colours—red and green—which the plate of Selenite exhibits when the two Nicols are parallel to or at right angles with one another, when combined, produce *white*. This can be immediately demonstrated by replacing the second Nicol with an ordinary crystal of Iceland spar (fig. 162, B), the principal plane of which is parallel to that of the first Nicol. It is traversed by both pairs of rays—those vibrating in PQ as well as those in MN—in consequence of which the former undergoes ordinary, the latter extraordinary refraction; two coloured images, the red and the green, are therefore now seen *at the same time* upon the screen, so placed, however, that they partially overlap. The part common to the two images when these colours are blended is *pure white*.

143. That the colours must be most lively when the principal plane of the lamina of Selenite forms an angle of 45° with the axis of the vibration of the polariser is easily demonstrated, for the two co-operating divisional rays are then equal in the intensity of their light, and the interference which gives rise to the colours is as complete as possible.

The reason that the colours observed in the crossed and parallel position of the Nicol are complementary to each other, is as follows. Let us suppose that a ray proceeding from the first Nicol strikes the anterior surface of the lamina in the point O (fig. 174), and communicates at a certain given moment to the particles of æther at O a motion in the direction OM, that is to say, upwards. Owing to the double refraction of the plate of Selenite placed at an angle of 45°, this motion is decomposed into two—of which the one is directed to the *right and upwards* (OR), the other to the *left and upwards* (OU). The former is decomposed into a motion *upwards* (OM), and into another to the *right* (OP); the second splits into a motion *upwards* and into one to the *left* (OQ). The two vertical part-motions thus, so far as only the action of the second Nicol comes into consideration, *coincide in direction*; the two horizontal ones are in direct opposition, or, in other words, the latter alone attain, owing to the decomposition effected by the polariscope, to a difference of path of a half wave-length, which is superadded to the difference of path already effected within the plate of Selenite. Were the Selenite plate just so thick that one ray lagged behind the other three half wave-lengths of the red (Fraunhofer's line, B), this colour must vanish when the Nicols are parallel; whilst the green (b), for the production of which a retardation of two whole wave-lengths occurs, attains its greatest brilliancy. The lamina therefore exhibits a green mixed colour when the Nicols are parallel. If the Nicols decussate, a half wave-length must be added to the difference of path of each kind of ray. The retardation of the red rays then amounts to two whole, that of the green to five half wave-lengths; and whilst the green rays extinguish each other, the red attain their highest brilliancy. The lamina therefore now appears of a red tint, which is exactly complementary to the green.

144. We can also obtain direct information respect-

ing the composition of the tint exhibited by a crystalline plate by effecting its decomposition with a prism. If, whilst the Selenite plate just described is introduced between the parallel Nicols, a prism be placed behind the second Nicol, a perfectly dark line appears in the red in the spectrum which is thrown upon the screen, proving that this colour is deficient in the green light which leaves the polariscope. If the second Nicol be now rotated, this stria, without altering its position, becomes progressively fainter, and ultimately, when the principal planes of the Nicols are inclined to each other at an angle of 45°, vanishes; for now, since only one of the two rays (RS or UV, fig. 174) penetrates the second Nicol, scarcely any interference takes place, and the white light, remaining undiminished in intensity, betrays itself by a spectrum without any spaces. As the Nicol is rotated still further, a slight shade makes its appearance in the green, which, as the Nicols approach to a position at right angles with one another, deepens into complete blackness.

The difference of path, and consequently also the tint of colour, dependent at any moment upon the prismatic decomposition, varies with the thickness of the plate. The thicker the Selenite plate is the greater is the number of dark striæ (fig. 153) that appear in the spectrum, and so much the nearer does its interference colour approximate to white, for reasons that have already been mentioned in speaking of the colours of thin plates. For a plate of Selenite consequently to exhibit lively colours, its thickness must not exceed 0·3 of a millimeter ($\frac{1}{72}$nd of an inch).

In order to exhibit at one and the same moment all the tints of colour that a plate of Selenite of every conceivable thickness may show, a wedge-shaped polished plate may be employed. By means of a polarising apparatus, with the arrangement of which (fig. 175) the reader is already familiar, the image produced by such a Selenite wedge may be thrown upon the screen; the colours, arranged in regular order parallel to the edge of the prism, exhibit the *same serial succession as in the Newtonian rings of colour,* and are therefore divided in the same manner into orders, and named in the same way (*see* § 118). The introduction of a concave and polished plate of Selenite resembling a concave lens into the polarising apparatus will even cause the colours to be arranged in concentric rings. It may be seen, in fact, that when the planes of vibration of the polarising apparatus are at right angles to one another, a system of coloured rings with dark central point makes its appearance, which differs from the Newtonian (fig. 151) rings only in the greater brilliancy of the colours.

It is unnecessary to mention that all the phenomena considered to be here represented to an audience upon a screen may also be observed by an individual if a Nörremberg's polarising apparatus be employed. When used for this purpose, a glass plate of about half its height is introduced into the apparatus (fig. 168, K', and 169, m), on which the crystal lamina to be examined is placed.

145. If two plates of Selenite of exactly the same thickness, and each of which by itself produces exactly the same tint, be now superimposed in such a manner that their principal planes coincide when introduced between the crossed Nicols, they exhibit another colour (fig. 173), namely, that which corresponds to a single plate of double the thickness of either alone. O plac-

ing the plates on one another in such a manner that their principal planes decussate at right angles, the screen will remain dark; nor does any tint of colour appear when the second prism is rotated, but the whole behaves just as if there were no plate of Selenite at all, for that ray, which travels more slowly in the first lamina, courses with greater rapidity in the second, its speed being just as much accelerated in this as it was retarded in the first. The two rays which leave the plate have therefore *no difference of path*, and cannot therefore give rise to any phenomena of interference of colour. Two unequally thick plates, crossed in the same way, act like a single plate the thickness of which is equal to the difference of thickness of the two plates, since the one only neutralises in part the action of the other. We may hence infer that interference colours may be produced by the decussation of two thick crystal plates neither of which appears coloured by itself, presupposing that the difference of their thickness is not too great.

This character may also be made use of in order to determine the gradation of the colour of the little plate of Selenite in the serial succession of the interference colours, with the aid of the wedge-shaped plate of Selenite; for if the plate of Selenite be placed in a cross position upon the wedge, it will be seen that the striæ are altered to just the extent that the plate covers the wedge. Along the line where the wedge is of the same thickness as the plate, this last abolishes the action of the wedge; at this spot therefore, when the Nicols are crossed, there must be a completely black line. The coloured stria, which in the uncovered part of the wedge forms the prolongation of the black line, now presents just that colour which the plate exhibits *per se*; and a glance is sufficient to show to which order this colour belongs.

146. In the above experiments, the polarised rays falling upon the crystal lamina have always been *parallel* to one another; in a plate of equal thickness throughout they have consequently to traverse paths of equal length, and their part-rays possess equal difference of path. A plate of equal thickness throughout exhibits, therefore, in *parallel* polarised light a single and uniform tone of colour in its whole extent.

To obtain a knowledge of the behaviour of crystalline plates in *converging* polarised light, a polarising

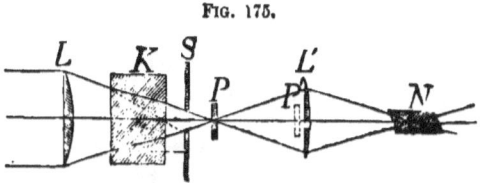

FIG. 175.

Polarising apparatus of Dubosq.

apparatus, constructed by Dubosq, is employed, the essential features of which are shown in fig. 175. The parallel rays of the sun falling on the lens, L, are collected into a cone which undergoes double refraction in a thick crystal of Iceland spar, K, which serves as a polariser. The cone of the ordinarily refracted rays, all of which vibrate at right angles to the principal cleavage plane of the Iceland spar, passes through the hole in the metal plate, S, whilst the cone of extraordinarily refracted rays are obstructed by the metal plate. The crystal plate, the action of which upon the converging polarised light is desired to be investigated, is placed at P, near the apex of the emerging cone of light; the rays diverging from the crystal plate, P, fall upon a second lens, which projects an image of the interference phenomena produced by the lamina upon a distant screen. Before the rays reach the screen, however, they are made to pass through the Nicol's prism, N, which serves as a polariscope.

147. The phenomena presented by plates of *uniaxial* crystals cut *at right angles to the optic axis* in convergingly polarised light is particularly worthy of note. That ray of the cone of light which strikes the plate vertically traverses it in the direction of the axis, and undergoes no double refraction; every other ray, however, undergoes double refraction, which is greater, because the path it has to traverse within the crystal is longer, in proportion as it strikes the crystal more and more obliquely. Thus it comes to pass that the differences of path are always greater the further the rays are distant from the axis of the cone of light; and since around and at an equal distance from the optic axis the two circumstances which determine the difference of path—the degree or amount of double refraction and the length of path—are equal, it follows that the same difference of path must exist for all points of a circle which may be conceived as drawn upon the screen around the point struck by the axial ray. A system of concentric rays consequently appears upon the screen, which exhibit a succession of colours similar to those in the rings of Newton.

When the planes of vibration of the polarising apparatus are crossed, the system of rings appears to be traversed by a *black cross* (fig. 176, A), the formation of which is easily explained; for since the optic axis is perpendicular to the surface of the crystal, every straight line, *MN*, *PQ*, *RS*, *UV* (fig. 174) drawn through the middle point of the system of rings upon the screen, corresponds to a principal plane. All rays that, proceeding from the polariser, strike upon the crystal-plate, vibrate parallel to *MN*, and consequently perpendicularly to *PQ*; they proceed therefore, without experiencing any decomposition, and with unaltered direction of vibration, both through the principal plane, *MN*, and through the principal plane, *PQ*—through the former by virtue of the extraordinary, and through the latter by virtue of the ordinary refraction—and are consequently not transmitted by the polariscope, the plane of vibration of which is placed at *PQ*. A black cross thus originates, the arms of which are parallel with the planes of the polarising apparatus. In every other principal plane, *RS*, making an angle with the plane of vibration, *MN*, of the polariser, a decomposition takes place into a ray vibrating in *RS*, and one perpendicular to this, the part-rays of which vibrating in *PQ*, in consequence of the prolonged difference of path, interfere, and thus give rise to the system of rings.

If the direction of vibration of the polariscope be parallel to that of the polariser, the rings that appear are complementary to the foregoing; and instead of the black cross, a *white one* (fig. 176, *B*) is obtained. After what has been already said, it is unnecessary to enter into any explanation of this phenomenon.

148. A plate of a biaxial crystal, the surfaces of which are perpendicular to the line which bisects the acute angle of the two optic axes—as for example a plate of Potassium nitrate—exhibits in the polarising apparatus, when the planes of vibration decussate, the beautiful phenomenon depicted in fig. 177. Two sys-

tems of rings are then seen, each of which surrounds an optic axis. The rings of higher order, approximating each other on the two sides, ultimately blend to form

FIG. 176.

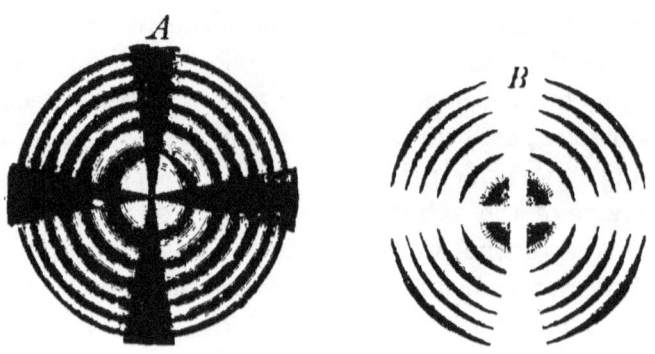

Rings of colour produced by uniaxial crystals.

peculiarly shaped curves, which, gently undulating, surround the two axial points. When the principal plane passing through the optic axes of the crystal plate coincides with one of the two planes of vibration

FIG. 177.

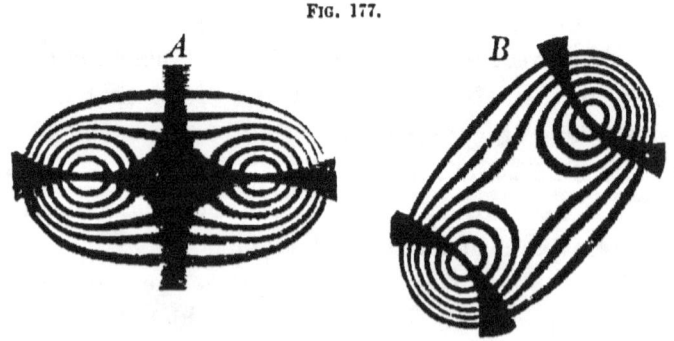

Rings of colour produced by biaxial crystals.

of the polarising apparatus, the double system of rings appears cut in two by a black cross (fig. 177, *A*); but if the crystal be rotated, the cross breaks up into two dark curved brushes, which, when the above-named principal plane forms an angle with the axis of polarisation of 45°, presents the appearance shown in fig. 177, B. If the polariser be rotated from the crossed into the parallel position, the rings present complementary colours to the foregoing, and the black brushes change to white ones. All these phenomena are explicable upon the laws of double refraction in biaxial crystals, and upon the same fundamental propositions on which the explanation of the coloured rings of uniaxial crystals rests.

The peculiar forms of the systems of rings affords a means of distinguishing biaxial from uniaxial crystals, by simple examination in a polarising apparatus. For the subjective observation of this phenomenon, the polarising apparatus of Nörremberg may be employed, a lens being added both above and below the glass plate (K', fig. 168) on which the crystal plate is placed.

The Tourmaline forceps or tongs (fig. 170) are still better adapted for this purpose, rendering the addition of the lenses unnecessary, since, when placed immediately in front of the eye, they permit the entry of rays into it coming from every direction.

149. It may be shown, with the aid of interference phenomena in polarised light, that singly refracting bodies like glass may also under certain circumstances become doubly refracting; that is to say, acquire the property of breaking up every incident ray of natural light into two polarised rays. If a square plate of glass fitted into a kind of vice be placed at the point P' (fig. 175) of Dubosq's polarising apparatus, and pressure be exerted upon it from above downwards by means of the screw, it is indeed compressed in this direction, but

extended in the horizontal one. The arrangement of its molecules is now no longer as before the same in all directions, and the plate becomes doubly refracting in consequence of the altered position of its molecules; and thus the screen, which previously to the pressure being exercised was dark on account of the crossed position of the planes of vibration, now presents a bright image of the plate, traversed by a dark cross. The property of double refraction may be permanently conferred upon a piece of glass by powerfully heating and then suddenly cooling it. If a disc of glass which has been thus treated be placed in the apparatus, a beautiful system of coloured rings with a black cross comes into view, just as in the case of a piece of Iceland spar cut at right angles to its optic axis. A black cross also appears in the case of a square glass plate, and in each of the four angles is a beautiful system of rings that may be compared with the eye of a peacock (fig. 178).

These phenomena furnish additional evidence of the intimate connection between the doubly refracting powers of different substances, and the arrangement of their molecules, to which reference has already been made in the chapter devoted to the double refraction of crystals. The double refraction of compressed and suddenly cooled glass is nevertheless essentially different from that of crystals. In order to project the system of rings of the glass disc upon the screen, it must be placed at the point P;* the rays by which it is struck are nearly

FIG. 178.

Polarisation image obtained from a suddenly cooled plate of glass.

* The little plate of gypsum, the Selenite wedge, and such bodies generally as are used in the experiments mentioned above, and the behaviour

INTERFERENCE OWING TO DOUBLE REFRACTION. 331

parallel, and traverse the plate in the same direction and with the same length of path. The difference of path which gives rise to the system of rings can therefore only be due to the fact that the *double refraction*, whilst the course of the rays remains unaltered, *increases towards the periphery of the plate*. In a crystal, on the contrary, the double refraction is at all points the same for the same direction of the rays.

of which in polarised light is desired to be investigated, must bo placed at the same point.

CHAPTER XXV.

CIRCULAR POLARISATION.

150. If a plate of Iceland spar cut at right angles to the optic axis be placed between two Nicol's prisms (fig. 179), the parallel polarised rays emerging from the first Nicol run collectively through the plate in the direction of the optic axis, without undergoing double refraction or any alteration in the direction of their vibration. On rotating the second Nicol, those variations of light and shade are only seen which would otherwise occur in the absence of the crystal plate.

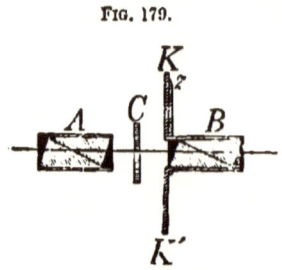

FIG. 179.

Two Nicol's prisms.

All uniaxial crystals, with the exception of Quartz, behave in the same way. If a polished Quartz plate, cut at right angles to the optic axis, be inserted between the two Nicols, the screen appears of a lively colour, the colour varying with the position of the Nicol, but never being dark. The colours, gradually passing into one another through all intermediate tints as the polariscope is turned, which are seen upon the screen, are successively red, orange, yellow, green, blue, violet; and these are repeated in the same order as the rotation is continued.

These colours are, however, by no means pure spectrum colours, and their composition, like the colours of Selenite, can be determined by prismatic decomposition. Thus, if the green light which is emerging from the polariscope in its present position be allowed to pass through a prism, a spectrum is produced the red part of which exhibits a perfectly black stria, whilst the orange and red are feebly, and the green and blue more vividly luminous. If the polariscope be turned in the same direction as before, the black line is seen to travel *gradually towards the more refrangible end of the spectrum*, and to blot out in succession the orange, yellow, green, blue, and violet colours, finally being lost in the extreme violet, in order to reappear at the red end of the spectrum. It is thus rendered evident that the tints which were seen when the prism was not used upon the screen are mixtures of all the simple colours left after the extinction of the one covered by the dark stria.

The position of the second Nicol, which corresponds to a definite position of the dark stria, is capable of being read off if the frame be provided with a marker, z, pointing to a divided circle, K, on the axis of which the tube rotates.

The Nicol can only extinguish those rays that vibrate at right angles to its principal plane. Before the Quartz plate was inserted, all vibrations were parallel to the vertically placed principal plane of the first Nicol (in the direction of the arrow, fig. 180); and they were therefore collectively extinguished and the screen was perfectly dark, since the principal plane of the second Nicol was horizontal, and thus decussated at right angles with that of the first. But after the Quartz plate is inserted (the thickness of which is 3·75 of a millimeter), the second Nicol must be rotated 60° from the crossed position, by which means the red rays undergo extinction in consequence of the dark stria in the red part of the spectrum. The direction of vibration of the red rays is consequently at right angles to the present position of the principal plane, and thus, through the action of the Quartz, *it has been rotated about* 60° from the vertical position which it previously had in common with all the other kinds of rays, and comes to occupy the position $r\,r'$ (fig. 180, upper figure). Similarly, the plane of vibration of the yellow rays has undergone a rotation of 90° ($g\,g'$), and that of the violet a rotation of 165° ($v\,v'$). In the adjoining figure the direction of the vibrations which are pursued by the chief colours of the spectrum, after their passage through the Quartz plate, is indicated in a very easily intelligible manner.

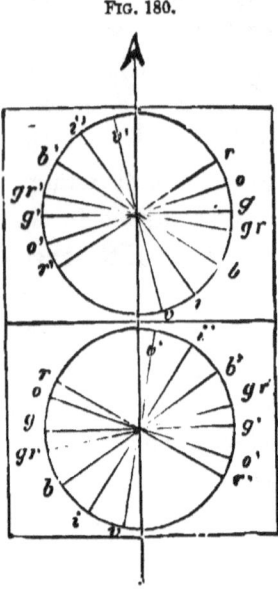

Fig. 180.

Rotation of the planes of vibration in Quartz.

The action of the Quartz plate thus consists in *effecting a rotation of the plane of vibration of the polarised rays, the amount of rotation varying for each kind of homogeneous light*, and being greater in proportion to the number of vibrations. In consequence of this dispersion of the colours in various directions of vibration, white light becomes broken up in a mode which is comparable with the dispersion of colour by ordinary refraction, and on this account has received the name of *circular* or *rotatory dispersion*.

The angle of rotation above given refers to a Quartz plate of 3·75 millimeters thick. When plates of various thickness are employed, it is found that for any given homogeneous colour the rotation increases in proportion to the thickness of the plate. If therefore the amount of rotation is known for any particular thickness, it may be immediately calculated for any other thickness. Broch measured the angle of rotation at which the dark stria in the spectrum occupied in succession the position of the principal Fraunhofer's lines, and found the following values for a Quartz plate of one millimeter in thickness:—

B	C	D	E	F	G
15° 30	17° 24	21° 67	27° 46	32° 50	42° 20.

151. In the case of the Quartz plate used in the foregoing experiments, whilst the dark line moves along the spectrum from the red to the violet end, the polariscope must be so rotated that the indicator, z, moves over the divided circle, K, in the direction of the hands of a watch, that is, to the *right*. But there are other specimens of Quartz in which the polariscope must be rotated in the opposite direction, or to the *left*, because the dark line moves in the spectrum from the violet to the red end. Quartz crystals are consequently distinguished as *rotating to the right or to the left*. Both kinds, with equal thickness of plate, rotate the plane of vibration of the same homogeneous light equally, but in opposite directions. The lower half of fig. 180 represents the rotation of the various colours in the case of a plate of 3·75 millimeters in thickness rotating to the left, just as the upper half shows it in the case of a plate of equal thickness, but rotating to the right.

152. In order to pave the way for the right understanding of the process by which the rotation of the plane of vibration is effected in a Quartz crystal, the motion must be investigated that is produced by the co-operation of two vibrations at right angles to each other; and for this purpose nothing is superior to the vibrations of an ordinary pendulum. A heavy leaden weight (fig. 181), pointed below, is suspended by a wire from the ceiling over a platter, the point when at rest being at O. Through the point O two lines, AB and CD, are drawn at right angles on the plane of the table. If the pendulum be brought to A, and then released, or if, when it is at rest, a blow be communicated to it in the direction OA, it swings to and fro in the line OA. In the same way it vibrates along the line CD if it be struck in this direction, or be brought to C or D and then released. The period of vibration, that is to say, the time requisite for its passage to and fro, is the same in whichever direction the vibrations are made to take place.

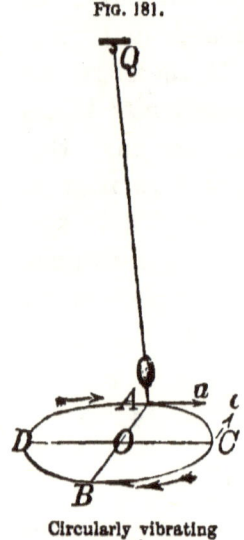

Fig. 181.

Circularly vibrating pendulum.

The question now arises, however, what movement will the pendulum perform if it be simultaneously acted upon by two impulses acting at right angles to each other? Let the pendulum be made to vibrate in the direction AB, and when it has reached the extreme point of its motion at A, let a blow be given to it in the direction Aa, at right angles to AB, the strength of which is just sufficient, if the pendulum be moving in this direction alone, to send it as far to the side from its present position as it was in the first instance moved at the moment of the blow from the position of rest at O. The result observed is that the lead weight describes with uniform velocity a circle, $ACBDA$, in the direction indicated by the arrows.

Had the vibration of the pendulum been measured from the moment in which it shortly before went in the direction BA through the point of rest, it would be found to have already performed a quarter-vibration* when it received the impulse in the direction Aa. It is thus seen *that two movements of vibration at right angles to each other, of which each is rectilinear in itself, combine to form a circular motion when one is a quarter-vibration before the other.* In the case illustrated by the figure, when the vibration directed to OA is antecedent to that directed to OC, the circular movement takes place in the direction of the hands of a watch, or to the *right*, as is indicated by the arrows. If the impulse be given in the opposite direction, a circular movement to the *left* is produced. The circular movement to the left is also engendered if the pendulum be first put into vibration in the direction OC; and when it has arrived at C, an impulse in the direction OA be given, that is, if the movement in the direction OA is a quarter-vibration behind that in OC. The time required for the completion of an entire circle is always equal to the period of vibration proper to the pendulum.

If the impulse given at A be more powerful than that which it originally received, the leaden weight is

---

* It may not perhaps be superfluous to observe that *by one entire vibration* is meant the motion $OAOBO$, or complete to and fro movement. The motion OA is consequently a quarter-vibration.

propelled to a greater distance laterally in the direction OC, and the pendulum moves in an ellipse the smaller axis of which is AB; but if the impulse be less powerful, AB becomes the greater axis of the ellipse described by the pendulum. Impulses applied to the pendulum whilst it is passing from O to A, or from O to B, likewise occasion elliptical paths of vibration, the axes of which however are no longer in the lines AB and CD.

If the lateral impulse in the direction OC be communicated at the moment when the pendulum passes through its position of rest, it assumes again a rectilinear movement, directed however neither towards A nor towards C, but along some intermediate line; in this case the one movement precedes the other either not at all or a certain number of half-vibrations.

153. The conditions of movement which were observed in the pendulum may also be followed in the case of light with the aid of thin crystalline laminæ. Mica, which *easily* splits up into still thinner plates than Selenite, is especially adapted for this purpose. If a thin plate of Mica be placed between the two Nicols (fig. 179), so that its principal plane RS (fig. 182), forms an angle of 45° with the axis of vibration, MN, of the polariser (the fig. 182 being now considered as applied to the surface of the lamina from which the light emerges), two equally luminous rays are found to emerge from the plate, of which one vibrates in RS, the other at right angles to it in UV. The particle of æther lying at O on the plane of emergence of the lamina is

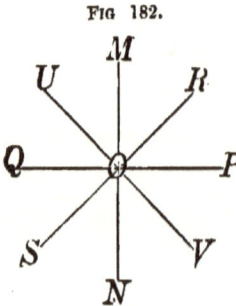

Fig. 182.

Decomposition of vibrations.

consequently, like the pendulum weight, affected contemporaneously by two impulses at right angles to each other, and assumes a circular, elliptic, or rectilinear motion according to the amount of the start which one vibration has over the other.

The Mica plate used in these experiments is just so thick that it occasions a difference of path of a quarter wave-length of yellow light between the two rays vibrating at right angles to each other. Under these circumstances it is obvious that for this colour the vibration of the more quickly propagated ray (which may be assumed to be UV), on arriving at the particle O precedes by a quarter-vibration that of the more slowly propagated ray (RS).

The particle O assumes therefore a circular movement to the right the period of revolution of which is equal to the duration of vibration of yellow light, and which communicates itself to the successive particles of æther arranged serially in the direction of the ray. Each of these moves in a circle, since its revolution begins somewhat later than the preceding, the plane of which is perpendicular to the ray around this; and if the coetaneous position of the æther particles at any moment be conceived to be connected by a curved line, a wavy line will be obtained which would wind round the ray like a screw, a complete turn of the screw corresponding to each wave-length.

A ray of light of this quality is said to be *circularly polarised*, whilst the rays that have hitherto been curtly termed 'polarised' will henceforward be referred to as *rectilinearly polarised*, because their vibrations take place in *straight lines* perpendicular to the direction of the ray.

The difference of path of the two rays vibrating at right angles to one another in the above-mentioned Mica plate amounts to an exact quarter-wave for the brightest yellow light alone; it is somewhat less for red rays and for blue somewhat more. The plate consequently communicates to the yellow rays alone a perfectly circular, whilst the rest have a more or less elliptic polarisation. Since, however, when the plate is thin the deviations from the circular form are very inconsiderable, the white light that is transmitted may be regarded as being almost completely circularly polarised.

154. The white fasciculus of rays proceeding from the quarter-wave Mica plate now demands examination. After allowing it to pass through the second Nicol, B, it will be found that the *screen remains equally bright in whatever direction the Nicol may be rotated*. A circularly polarised ray may in fact, since its quality is the same all round, exhibit no laterality; it behaves itself when examined with a Nicol like an ordinary ray of light. That it is not such a natural ray is immediately rendered apparent if a second Mica plate of equal thickness, but with its principal plane at right angles, be interposed. The original rectilinear polarisation is again shown to be present; the screen ceases to be illuminated when the plane of vibration of the second Nicol decussates with that of the first. The very case mentioned above in regard to the pendulum is before us, namely that neither of the two perpendicular vibrations precedes the other, so that the two equal vibrations, OR and OU, combine to produce a rectilinear vibration, OM, the axis of which bisects the angle, ROU. If the second Mica plate be superimposed upon the first, with its principal plane parallel, the dif-

ference of path between OU and OR amounts to a half wave-length, and again gives rise to a rectilinearly polarised ray which now vibrates in PQ, and consequently disappears when the plane of vibration of the second Nicol is parallel to that of the first. A quarter-wave Mica plate may thus be used for the purpose of recognising circularly polarised from rectilinearly polarised and from natural light, as it is capable of converting a rectilinearly polarised into a circularly polarised ray of light; it may also, conversely, change circularly polarised light into rectilinearly polarised, whilst it allows a natural ray of light to continue unaltered.

155. In the above-mentioned experiment with a circularly polarising Mica plate, it has been taken for granted that the more rapidly moving ray vibrates in the axis OU; on this supposition the circular movement of the æther particles takes place to the right. If the Mica plate be rotated in its plane 90°, so that the vibration in the axis OR is accelerated about a quarter-vibration, the plate occasions the light to be polarised circularly to the left. When this is examined with the Nicol and with the second Mica plate, it behaves in exactly the same manner as that polarised to the right, and cannot be distinguished from it by these means. The difference, however, can be instantly recognised if a plate of Selenite, with its principal plane placed at 45°, be interposed between the Mica plate, C, and the second Nicol, B (fig. 179), at right angles with the first, the phenomena of colour of which in rectilinearly polarised light are now sufficiently known. The light upon the screen now appears coloured, *the colour varying* according to whether the Mica plate is introduced in right- or in left-handed circular polarisation.

If, for example, the colour be in the first instance bluish green, the complementary rose-red tint appears in the second instance. In that case the principal planes of the Selenite and of the Mica plate are parallel to each other, and to the difference of path which the Selenite occasions must be added that difference, amounting to a quarter wave, which is induced by the Mica plate; in the second case, where the principal planes of the two plates decussate at right angles to each other, the difference of path occasioned by the plate of Selenite is diminished by a quarter wave. The difference of path in light polarised circularly to the right exceeds consequently by a half wave that polarised circularly to the left, so that there all those rays are extinguished which are here most brilliant, and *vice versâ*. The mixed colours therefore which occur in the two cases must be complementary to each other.

156. Recurring for a moment to the pendulum (fig. 181), and conceiving that the leaden weight whilst it is at A (fig. 183) receives an impulse not only in the direction $A a$, but coincidently also an equally powerful impulse in the opposite direction, $A a'$, the first impulse, combined with the impulse which the pendulum already possesses in the direction of the line $A B$, would lead to a circular movement to the right; the second, to a similar movement to the left. If the two impulses acted simultaneously, they would neutralise each other, and the pendulum would continue to vibrate to and fro along the straight line, $A B$, as if nothing had happened.

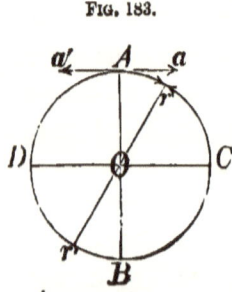

Fig. 183.

Combined effect of two opposite circular vibrations.

But supposing the second impulse to occur later, after the pendulum had in consequence of the first impulse already performed the circular movement, Ar, and supposing this impulse to be in opposition to the direction of the movement it possesses at the point r, a rectilinear movement will obviously be developed along rr'. From this it results that a vibrating body acted on coincidently by two equal but opposite circular forces will acquire a rectilinear vibrating movement, which takes place along that diameter of the circle at the terminal point of which it received the impulses.

If this proposition be applied to the vibrations of light, it follows *that a rectilinearly polarised ray is always the result of the combined effect of two rays of light polarised circularly in opposite directions, of equal brilliancy and equal number of vibrations, following the same path*; and conversely, it may be said *that every rectilinearly polarised ray may be regarded as composed of two equally bright rays of light polarised circularly in opposite directions*.

157. This representation or explanation of the phenomena founded on the general laws of motion, and to the effect that a rectilinearly polarised ray of light consists of two rays polarised circularly in opposite directions, would only possess a theoretic significance were there not bodies which act upon light polarised circularly to the right differently to light polarised circularly to the left. Fresnel has shown that Quartz is such a body.

The fact of the rotation of the plane of vibration through a plate of Quartz becomes perfectly intelligible if it be admitted that rays polarised circularly in opposite directions are propagated with different velocities along the axis of a crystal of Quartz. A rectilinearly polarised ray of light must, on its entrance into a Quartz plate, be broken up into two rays polarised circularly in opposite directions, which, after they have traversed the plate with unequal velocity, on their exit again combine to form a rectilinearly polarised ray, the plane of vibration of which differs either to the right or left of that of the incident ray according as the right or left circular impulse is antecedent and affects earlier the particles of æther in contact with the surface of emergence. The greater the thickness of the Quartz plate, the greater is the retardation of one of the two rays, and the greater must be the rotation of the plane of vibration. The circumstance that equally thick plates of Quartz rotate the plane of vibration to the right and to the left *to the same extent*, although in opposite directions, indicates that the rapidity of propagation of the rays polarised circularly in opposite directions are the same in the two kinds of Quartz, and are only interchangeable so far that that ray which has a greater velocity in the one crystal moves more slowly in the other.

158. If the two kinds of circularly polarised rays are propagated with different velocities parallel to the axis of the Quartz, a peculiar kind of *double refraction* must take place in this direction, by means of which an incident rectilinearly polarised ray is decomposed into two rays polarised circularly in opposite directions. In the Quartz plates that have hitherto been employed, and which were struck rectilinearly by the incident rays, an actual

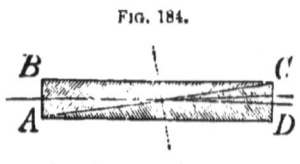

Fig. 184.

Double prism of Quartz.

decomposition can certainly not take place, because although the two rays are propagated with different velocities, they course in the same direction. Fresnel, by an ingenious combination of two prisms of Quartz rotating in opposite directions, did however effect this decomposition, and thus demonstrated beyond a doubt the correctness of the explanation previously given of the rotation of the plane of vibration.

Fresnel's *double prism* (fig. 184) consists of two elongated rectangular prisms of Quartz each having an acute angle ACB of 7°, one of which is cut from a prism rotating to the right, and the other from a prism rotating to the left. Being cemented together by their oblique surfaces, AC, they form a rectangular column the terminal surfaces of which, AB and CD, are perpendicular to the optic axis. If a rectilinearly polarised beam be allowed to fall through a round opening upon the surface AB, it undergoes decomposition into two rays polarised circularly in opposite directions which traverse the first prism with different velocities, but in a path common to both. The ray which in the first prism was the most rapid, on entering the second prism becomes the less rapid of the two, and therefore approaches to the perpendicular (indicated in the figure by the dotted line); on the other hand, the ray moving more slowly in the first prism is propagated more rapidly in the second, and must consequently recede from the perpendicular. Two separate fasciculi consequently emerge from the surface CD, which produce two round spots of light upon the screen, the borders of which overlap to some extent. When looked at through a Nicol placed between CD and the screen, the two beams prove to be circularly polarised, and if a plate of Selenite be placed between the double prism and the Nicol at an angle of 45°, one of the spots of light appears of a bluish green, the other of a rose-red tint, whilst the area common to both remains white. The occurrence of these complementary colours demonstrates that one of the beams is circularly polarised to the right, the other to the left. This experiment therefore furnishes decisive proof that a rectilinearly polarised ray of light is decomposed by the Quartz into two rays moving with unequal velocity and polarised circularly in opposite directions.

159. The power of circular double refraction belongs to only a few substances besides quartz, and is not associated with any definite crystalline system; it is exhibited by a few singly refracting crystals belonging to the regular system, as for example by Sodium chloride *in all directions*. In doubly refracting crystals, as for example in Quartz, it can only be perceived in directions that are nearly parallel to the optic axis, because in every other direction they are concealed by the ordinary double refraction.

Circular double refraction consequently appears not to be dependent upon any special arrangement of the molecules, but rather upon a peculiar structure of the molecules themselves, which may no doubt betray itself in crystalline bodies by the external form of the crystal, as in fact is the case with Quartz. This opinion is materially supported by the fact, *that many fluids possess the power of effecting double circular refraction, and consequently the power of rotating the plane of vibration of rectilinearly polarised light.*

The plane of vibration is rotated to the right by aqueous solutions of cane- and grape-sugar, tartaric acid,

CIRCULAR POLARISATION. 347

oil of lemons, and by an alcoholic solution of camphor. It is rotated to the left by oil of turpentine, by cherry-laurel water, and by solution of gum arabic.

As the rotatory power of these fluids is very inferior to that of quartz, it is necessary in order to observe it conveniently to employ layers of considerable thick-

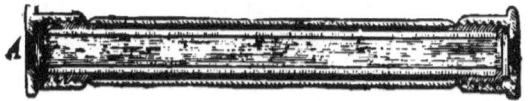

Tube for the reception of circularly polarising fluids.

ness, which is best accomplished by filling tubes with them, the ends of which are closed with plane glass plates (fig. 185).

If such a tube, filled with solution of sugar, be placed between the crossed Nicols, the previously dark screen immediately becomes illuminated, and from the amount of rotation which must be communicated to the polariscope, in order that the screen may again be darkened, the angle may be known which the solution of sugar has rotated the plane of vibration of the incident rectilinearly polarised light. This rotation is proportional on the one hand to the thickness of the layer, and on the other to the amount of active substance (sugar) contained in the fluid, and as it is known that with a tube 20 centimeters (7·8 inches) in length, the rotation of the plane of vibration amounts to $1°·333$ for each gramme (15·44 grains) of sugar contained in 100 cubic centimeters (6·102705 cubic inches, or rather less than one-sixth of a pint) of the solution, the amount of sugar contained in the solution may be immediately determined from the amount of rotation produced by the solution.

160. In order to attain the greatest accuracy in the determination of the amount of sugar contained in the solution, an instrument is desirable which renders a very small rotation perceptible. Such an instrument is found in the *double quartz plate* (fig. 186) first con‑

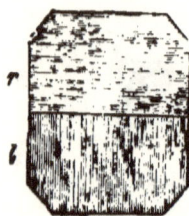

Fig. 186.

Double plate composed of right and left rotating Quartz.

structed by Soleil. It is composed of two quartz plates, cut at right angles to the axis and cemented together, of which one rotates to the right and the other to the left, whilst each has a thickness of 3·75 millimeters. If now a double plate of this kind be placed between the two Nicols the planes of vibrations of which are parallel, and if the image be cast by means of a lens upon a screen, both halves of the plate will be found to exhibit the same violet tint of colour. On the interposition of the tube filled with the solution of sugar, a dissimilarity of colour is immediately observed in the two halves of the plate, one half presenting a bluish, the other a reddish tint. The plane of vibration of each colour contained in white light is rotated to an equal amount in each half of the double plate, but in the one half the rotation is to the right and in the other to the left, as has been indicated in the corresponding halves of fig. 180. If the principal planes of the Nicol be parallel to each other (in the direction of the arrow) the two halves must exhibit the same tint of colour. A glance at the figure above alluded to suffices to show that in this position of the Nicol the yellow disappears, and that consequently a violet colour must appear as a result of the mixture of the remaining colours.

As the solution of sugar rotates the planes of vibra‑

CIRCULAR POLARISATION. 349

tion of all rays to the right, the rotation is increased in the half rotating to the right and diminished in the half rotating to the left; in the former, therefore, the planes of vibration of the orange tints, in the latter those of the green rays, appear in the position previously occupied by the planes of vibration of the yellow rays. The former half will therefore exhibit a blue, the latter a red tone of colour. In order to ascertain how much the solution of sugar has rotated the plane of vibration, it is only requisite to rotate the second Nicol till the two halves of the plates again appear of the same colour.

161. As the rapid and convenient determination of the amount of sugar contained in a saccharine solution is of great practical importance in an economical point of view, an apparatus has been constructed with this object in view, called a Saccharimeter.

The Saccharimeter of Soleil has (fig. 187) the previously described double plate at r between the two Nicol's prisms S and T, the planes of vibration of which are fixed parallel to each other. The change of colour which the tube filled with solution of sugar introduced at m induces is, however, not compensated for by rotating the polariscope, T, but by a highly ingenious compensating arrangement placed at e (*the compensator*).

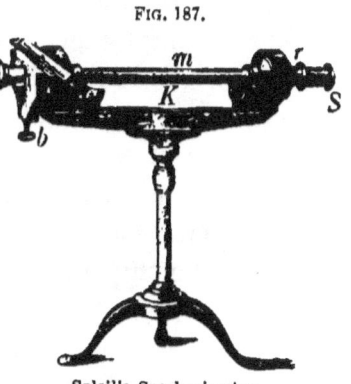

FIG. 187.

Soleil's Saccharimeter.

The rays emerging from m pass first through a quartz plate rotating to the right, cut at right angles to the axis, and then through two wedges, r and o, cut from a quartz plate rotating to the left (fig. 188), and which by means of a screw, *b*, can be moved towards each other.

Fig. 188.

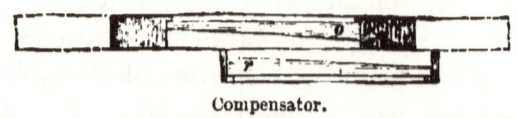

Compensator.

When in contact they form a quartz plate, cut perpendicularly to the axis, which is of the same thickness as the first-mentioned one, and therefore completely neutralises its rotation to the right. If they are moved from this position to either side, the extent which the rays have to traverse in the two wedges together is augmented or diminished; the two wedges together thus form a quartz plate rotating to the left, the thickness of which within certain limits can be varied at will and can be made equal to, or larger or smaller than that of the quartz plate rotating to the right. The alteration of thickness in each movement of the screw can be read off by means of the indicator, *v*, upon a small scale, *e*, to the 1000th of a millimeter. According as the rotation of the plate to the right, or the rotation of the system of wedges to the left, is allowed to predominate, the action of the compensator is equivalent with that of a plate of quartz rotating to the right or to the left, the thickness of which may be exactly determined.

In order to compensate the difference of colour between the two halves of the double plate, which is brought about by virtue of the rotation to the right of the solution of sugar, the compensator must be arranged for an equal amount of left-handed rotation; then, by reading the scale, *the thickness is obtained of a quartz plate*

CIRCULAR POLARISATION. 351

which possesses the same power of rotation as the saccharine solution under examination. And as it has been ascertained by carefully made experiments that a solution of sugar which contains 16·35 grammes (252·44 grains) of pure crystallised sugar in 100 cubic centimeters exerts as great a rotating power in a tube 20 centimeters in length as a quartz plate 1 millimeter in thickness, it is only necessary to multiply the number read off upon the scale by 16·35 in order to know the weight of sugar contained in 100 cubic centimeters of the solution.

And now, in conclusion, let a brief retrospective glance be cast upon the subjects that have here been treated of. The reply to the question, What is Light? was the end in view. Proceeding step by step by the light of experience, the various phenomena of light were considered, the laws investigated to which those phenomena are subject, and the useful applications which life and science have made from them. Finally, a fact was disclosed (Fresnel's interference experiment) which pressed home to us the conviction that light must consist in the undulatory movement of an attenuated elastic substance. Having arrived at this stand-point, it was requisite to call a halt in order to reconsider the phenomena already observed, and when it had been ascertained that the previously isolated facts became in succession, under this point of view, united into one whole, further advances were made, and new facts obtained which threw additional light upon the nature and essence of light. The phenomena of polarisation demonstrated, in point of fact, *that the vibrations of light take place at right angles to the direction of the rays.* The last part of this work gave results that did not at first appear to be capable of useful application to the life of man until quite recently, when an apparatus has been constructed of pre-eminent practical importance.

It is the task of science to strive after truth without having any secondary object in view. If it remain true to this ideal, the practical applications will fall into its lap as the ripe fruits of knowledge.

INDEX.

ABS

ABSORPTION lines, 172
— of light, 242
— spectra, 173
Achromatic lens, 141, 146
Achromatism, 134
Æsculin, fluorescence of, 183
Æther, 213
Alcohol, index of refraction of, 60
Angles of incidence and refraction, 57
Ångström on wave-lengths, 270
Anthracite, index of refraction of, 312
Arc of flame, Volta's, 9

BARIUM, spectrum analysis of, 150
Bartholinus on double refraction, 282
Becquerel's phosphoriscope, 191
Becquerel on wave-length of ultra red rays, 280
Biot's polarising apparatus, 307
Black cross of polarised light, formation of, 326
— flame, 163
Blood, absorption spectrum of, 174
Bunsen's burner, 3
— apparatus for absorbing Sodium light, 162
— photometer, 24
— spectroscope, 148

CÆSIUM, spectrum analysis of, 152
Calcium fluoride, 184
— spectrum analysis of, 150
Camera obscura, 19, 101

DIA

Canada balsam, index of refraction of, 60
Carbonic disulphide, index of refraction of, 60
Carboniferous strata, their relation to solar energy, 256
Chlorophyll, absorption spectrum of, 174
Christiansen on anomalous dispersion of light, 244
Circular polarisation of light, 332
Cobalt, absorption spectrum of, 175
Colours, complementary, 120
— dispersion of, 112
Compensator, 350
Concave lenses, 87
— mirrors, 40
Conjugate foci, 42
— points, 81
Convex lenses, 79
— mirrors, 49
Copper, absorption spectrum of, 176
Critical angle, 63
Cross, black, of polarised light, formation of, 326
Crown glass, index of refraction of, 60

DARK rays of the solar spectrum, 203
Deflection, minimum of prisms, 70
— without dispersion, 136
Determination of conjugate points, 90
— of the focal distance of lenses, 90
Diamond, index of refraction of, 60
— critical angle of, 63

DIA

Diathermancy of rock salt, 203
Dichroism, 313
Didymium, absorption spectrum of, 177
Diffraction apparatus, 260
— of light, 258
Dispersion of colour, 112, 140, 242
— without deflection, 137
— of light, 242
— anomalous, of light, 244
Doppler on pitch of sound and tone of colour, 245
Double prism of Fresnel, 345
Double refraction, 282
Drummond's lime-light, 7
Dubosq's lamp, 95
— polariser, 325
— regulator, 10
Dutch telescope, 107

ECLIPSES, cause of. 17
Electric lamp, 8
Enlarged images, 46, 87
Erbium, absorption spectrum of, 177
Esselbach on wave-lengths of ultra red rays, 280
Extraordinarily refracted rays, 283
Eye, general construction of, 102

FILMS, colours of, 273
Flames of candles and lamps, 4
Flint-glass, index of refraction of, 60
Fluor spar, 184
Fluorescence, 183
Foci of concave mirrors, 41
— conjugate, 42
— of lenses, 80
Fresnel's double prism, 345
— mirror experiment, 207
Front view telescope, 110
Fuchsin, anomalous dispersion power of, 243

GALILEO, telescope of, 107
Galvanometer, 199
Gases, spectra of, 155
— — inversion of, 164
Geissler's tubes, 155, 187, 188

LEN

Ghost experiment, 32
Glass, critical angle of, 63
Goniometer, 34
Grating spectrum, 266
Gregory's telescope, 110

HEAT, action of, 197
— measurement of, 199
— curve of, in spectrum, 201
Heliostat, 32
Herschel's telescope, 109
Hooke's theory of light, 229
Huggins' estimate of rate of movement of Sirius, 246
Huyghens' principle, 229

ICELAND spar, double refraction of, 284
Illumination, law of, in regard to distance, 23
— aerial, 47
— apparent, 47
Images, inverted, 46
— virtual, 47
Incandescence, 2
Indium, refraction of light of, 114
— spectrum analysis of, 153
Induction apparatus, 154
Interference of light, 316
— of sonorous waves, 212
Inverted images, 46, 86
Iodine, absorption spectrum of, 173

KALEIDOSCOPE, 36
Kepler's telescope, 104
Kundt on anomalous dispersion of light, 244

LAMINÆ, colours of thin, 273
Law of illumination in proportion to distance, 22
— reflection of light, 28
Lamp, Drummond's, 7
— Dubosq's, 95
— electric, 8
— magnesium, 7
— oxygen, 5
— petroleum, 5
Lantern, magic, 97
Lenses, 78

INDEX.

LEN

Lenses, axis of, 79
— bi-convex, 78
— bi-concave, 78
— centre of curvature of, 79
— convexo-concave, 78
— concavo-convex, 78
— foci of, 80
— plano-convex, 78
— plano-concave, 78
Light, absorption of, 242, 253
— dispersion of, 112, 224
— rays of, 14
— rectilinear propagation of, 14
Lime-light, 7
Lithium, refraction of light of, 114
— spectrum analysis of, 150
Litmus, absorption spectrum of, 178
Lunar eclipses, causes of, 17

MAGIC LANTERN, 37
Magnesium lamp, 7
Malus on polarisation of light by reflexion, 311
Menilite, index of refraction of, 312
Meyerstein's spectrometer, 144
Mica, reflected light from plate of, 279
Microscope, compound, 103
— simple, 102
— solar, 98
Minimum deflection of prisms, 70
Mirror experiment, Fresnel's, 208
Mirror sextant, 37
Mirrors, concave, 40
— convex, 49
Motion, modes of propagation of, 210
Müller on wave-lengths of ultra red rays, 281
Müller's lines, 280

NAPHTHALIN, red fluorescence of, 189
Newton's colour glass, 273
— rings, 274
— telescope, 109
Nicol's prisms, 304
Nitrous oxide, absorption spectrum of, 173
Nörremberg's polarising apparatus, 308
Nucleus of shadows, 16

RAY

OBJECTIVE, 103
Ocular, 103
Optical instruments, 95
Ordinarily refracted rays, 283
Origin of light, 248
Oxygen lamp, 5

PARALLAX of sun, 18
Pennine, dichroism of, 313
Penumbra, nature of, 16
Permanganate of potash, absorption spectrum of, 174
Petroleum lamp, 5
Photography, principles of, 194
Photographic action of solar spectrum; curve of, 205
Photometer, Bunsen's, 24
— Rumford's, 24
Phosphorescence, 183, 192
Phosphoriscope, 191
Plates, colours of thin, 273
— bichromate, 176
Polarisation of light, 293
— circular, of light, 332
Polarising apparatus, 303
— Biot's, 307
— Dubosq's, 325
— Nörremberg's, 309
Potassium, spectrum analysis of, 150
Principal axis, 40
— comparing, 160
— hollow, 71
— minimum deflection of, 70
Principle, Doppler's, 245
— of conservation of energy, 253
— of interference, 217
— Huyghens', 229
Prism, double, of Fresnel, 345
Prisms, Nicol's, 304
— refracting angle of, 68

QUARTZ crystals, circular polarisation of light by, 335
Quinia, fluorescence of, 187

RAINBOW, mode of formation of, 122
Rays of light, 14

INDEX.

REA

Real images, 81
Reflecting goniometer, 34
— telescope, 110
Reflexion, polarisation of light by, 306
Refraction, 56
— angle of, 57
— index of, 60
Refractors, 105
Regulator of Dubosq, 10
Resonance, 252
Reusch's heliostat, 33
Rock salt, diathermancy of, 202
Rose de Magdala, fluorescence of, 189
Rosse's, Lord, telescope, 109
Rubidium, spectrum analysis of, 152
Rumford's photometer, 24

SACCHARIMETER of Soleil, 349
Selenite, colours exhibited by, 328
— interference, experiments with plates of, 319
Sextant, 37
Shadows, nature of, 15
— nucleus of, 16
— penumbra of, 16
Sirius, rate of movement of, 247
Sodium, refraction of light of, 114
— spectrum analysis of, 150
Solar eclipses, cause of, 17
— microscope, 98
— spectrum, general view of, 205
— — length of, 281
Soleil's saccharimeter, 349
Sound, propagation of, 211
— interference of waves of, 212
Spar, Iceland, double refraction of, 284
Spectra of gases, 155
Spectrometer, 144
Spectroscope, Browning's, 149
— Bunsen's, 148
— direct vision, 149
— Hoffman's, 149

WAV

Spectroscope, dispersing, 169
Spectrum, analysis, 149
— calorific action of, 201
— continuous, 118, 250
— interrupted, 157
— nature of, 117
— solar, 158
— thermotic curve of, 201
Spherical mirrors, 40
Strontium, light of, 116
— sulphide, phosphorescence of, 192
Sun, spectrum analysis of, 159, 165

TELESCOPE, Galileo's, 107
Gregory's, 110
— Kepler's, 104
— Newton's, 109
— Refractors, 105
Terbium, absorption spectrum of, 177
Thallium, refraction of light of, 114
— spectrum analysis of, 153
Theodolite, 106
Theory, Huyghens', 229
Thermopile, 199
Thermotic curve of the spectrum, 201
Tourmaline forceps or tongs, 314

ULTRA red rays, 280
Undulations of sound, 211
— of water, 213
Undulatory motion, 210
Uniaxial crystals, rings of colour produced by, 328
Uranium, fluorescence of, 187

VIRTUAL images, 47, 50, 87, 89
Volta's arc of flame, 9

WATER, critical angle of, 63
Index of refraction of, 60
Wave-rays, 215

D. APPLETON & CO.'S PUBLICATIONS.

THE WARFARE OF SCIENCE WITH THE-OLOGY. A History of the Warfare of Science with Theology in Christendom. By ANDREW D. WHITE, LL. D., late President and Professor of History at Cornell University. In two volumes. 8vo. Cloth, $5.00.

"The story of the struggle of searchers after truth with the organized forces of ignorance, bigotry, and superstition is the most inspiring chapter in the whole history of mankind That story has never been better told than by the ex-President of Cornell University in these two volumes. . . . A wonderful story it is that he tells."—*London Daily Chronicle.*

"A literary event of prime importance is the appearance of 'A History of the Warfare of Science with Theology in Christendom.'"—*Philadelphia Press.*

"Such an honest and thorough treatment of the subject in all its bearings that it will carry weight and be accepted as an authority in tracing the process by which the scientific method has come to be supreme in modern thought and life."—*Boston Herald.*

"A great work of a great man upon great subjects, and will always be a religio-scientific classic."—*Chicago Evening Post.*

"It is graphic, lucid, even-tempered—never bitter nor vindictive. No student of human progress should fail to read these volumes. While they have about them the fascination of a well-told tale, they are also crowded with the facts of history that have had a tremendous bearing upon the development of the race."—*Brooklyn Eagle.*

"The same liberal spirit that marked his public life is seen in the pages of his book, giving it a zest and interest that can not fail to secure for it hearty commendation and honest praise."—*Philadelphia Public Ledger.*

"A conscientious summary of the body of learning to which it relates accumulated during long years of research. . . . A monument of industry."—*N. Y. Evening Post.*

"A work which constitutes in many ways the most instructive review that has ever been written of the evolution of human knowledge in its conflict with dogmatic belief. . . . As a contribution to the literature of liberal thought, the book is one the importance of which can not be easily overrated."—*Boston Beacon.*

"The most valuable contribution that has yet been made to the history of the conflicts between the theologians and the scientists."—*Buffalo Commercial.*

"Undoubtedly the most exhaustive treatise which has been written on this subject. . . . Able, scholarly, critical, impartial in tone and exhaustive in treatment."—*Boston Advertiser.*

New York: D. APPLETON & CO., 72 Fifth Avenue.

D. APPLETON & CO.'S PUBLICATIONS.

THE SCIENCE OF LAW. By Professor SHELDON AMOS, M.A. 12mo. Cloth, $1.75.

CONTENTS.—Chapter I. Recent History and Present Condition of the Science of Law; II. Province and Limits of the Science of Law; III. Law and Morality; IV. The Growth of Law; V. The Growth of Law (continued); VI. Elementary Conceptions and Terms; VII. Law in Relation to (1) the State, (2) the Family, (3) the other Constituent Elements of the Race; VIII. Laws of Ownership of Property; IX. Law of Contract; X. Criminal Law and Procedure; XI. The Law of Civil Procedure; XII. International Law; XIII. Codification; XIV. Law and Government.

"Professor Amos has certainly done much to clear the science of law from the technical obscurities which darken it to minds which have had no legal training, and to make clear to his 'lay' readers in how true and high a sense it can assert its right to be considered a science, and not a mere practice."—*Christian Register.*

THE SCIENCE OF POLITICS. By Professor SHELDON AMOS, M.A., author of "The Science of Law," etc. 12mo. Cloth, $1.75.

CONTENTS.—Chapter I. Nature and Limits of the Science of Politics; II. Political Terms; III. Political Reasoning; IV. The Geographical Area of Modern Politics; V. The Primary Elements of Political Life and Action; VI Constitutions; VII. Local Government; VIII. The Government of Dependencies; IX. Foreign Relations; X. The Province of Government; XI. Revolutions in States; XII. Right and Wrong in Politics.

"The author traces the subject from Plato and Aristotle in Greece, and Cicero in Rome, to the modern schools in the English field, not slighting the teachings of the American Revolution or the lessons of the French Revolution of 1793. Forms of government, political terms, the relation of law, written and unwritten, to the subject, a codification from Justinian to Napoleon in France and Field in America, are treated as parts of the subject in hand. Necessarily the subjects of executive and legislative authority, police, liquor, and land laws are considered, and the question ever growing in importance in all countries, the relations of corporations to the state."—*N. Y. Observer.*

DIGEST OF THE LAWS, CUSTOMS, MANNERS, AND INSTITUTIONS OF ANCIENT AND MODERN NATIONS. By THOMAS DEW, late President of the College of William and Mary. 8vo. Cloth, $2.00.

No pains have been spared by the author to secure accuracy in facts and figures; and in doubtful cases references are given in parentheses, so that the student can readily satisfy himself by going to original sources. The department of Modern History, too often neglected in works of this kind, has received special care and attention.

ROMAN LAW; Its History and System of Private Law. In Twelve Academical Lectures. By Professor JAMES HADLEY. 12mo. Cloth, $1.25.

"The most valuable short account of the nature and importance of the body of Roman law. The lectures are free from embarrassing technical details, while at the same time they are sufficiently elaborate to give a definite idea of the nature and the greatness of the subject."—*Dr. C. K. Adams's Manual of Historical Literature.*

New York: D. APPLETON & CO., 72 Fifth Avenue.

D. APPLETON & CO.'S PUBLICATIONS.

WORKS BY ARABELLA B. BUCKLEY (MRS. FISHER).

THE FAIRY-LAND OF SCIENCE. With 74 Illustrations. 12mo. Cloth, gilt, $1.50.

"Deserves to take a permanent place in the literature of youth."—*London Times.*

"So interesting that, having once opened the book, we do not know how to leave off reading."—*Saturday Review.*

THROUGH MAGIC GLASSES, *and other Lectures.* A Sequel to "The Fairy-Land of Science." Illustrated. 12mo. Cloth, $1.50.

CONTENTS.

The Magician's Chamber by Moonlight.	An Hour with the Sun.
Magic Glasses and How to Use Them.	An Evening with the Stars.
Fairy Rings and How They are Made.	Little Beings from a Miniature Ocean.
The Life-History of Lichens and Mosses.	The Dartmoor Ponies.
The History of a Lava-Stream.	The Magician's Dream of Ancient Days.

LIFE AND HER CHILDREN: *Glimpses of Animal Life from the Amœba to the Insects.* With over 100 Illustrations. 12mo. Cloth, gilt, $1.50.

"The work forms a charming introduction to the study of zoölogy—the science of living things—which, we trust, will find its way into many hands."—*Nature.*

WINNERS IN LIFE'S RACE; *or, The Great Backboned Family.* With numerous Illustrations. 12mo. Cloth, gilt, $1.50.

"We can conceive of no better gift-book than this volume. Miss Buckley has spared no pains to incorporate in her book the latest results of scientific research. The illustrations in the book deserve the highest praise—they are numerous, accurate, and striking."—*Spectator.*

A SHORT HISTORY OF NATURAL SCIENCE; *and of the Progress of Discovery from the Time of the Greeks to the Present Time.* New edition, revised and rearranged. With 77 Illustrations. 12mo. Cloth, $2.00.

"The work, though mainly intended for children and young persons, may be most advantageously read by many persons of riper age, and may serve to implant in their minds a fuller and clearer conception of 'the promises, the achievements, and the claims of science.'"—*Journal of Science.*

MORAL TEACHINGS OF SCIENCE. 12mo. Cloth, 75 cents.

"A little book that proves, with excellent clearness and force, how many and striking are the moral lessons suggested by the study of the life history of the plant or bird, beast or insect."—*London Saturday Review.*

New York: D. APPLETON & CO., 72 Fifth Avenue.

D. APPLETON & CO.'S PUBLICATIONS.

RICHARD A. PROCTOR'S WORKS.

OTHER WORLDS THAN OURS: *The Plurality of Worlds, Studied under the Light of Recent Scientific Researches.* By RICHARD ANTHONY PROCTOR. With Illustrations, some colored. 12mo. Cloth, $1.75.

CONTENTS.—Introduction.—What the Earth teaches us.—What we learn from the Sun.—The Inferior Planets.—Mars, the Miniature of our Earth.—Jupiter, the Giant of the Solar System.—Saturn, the Ringed World.—Uranus and Neptune, the Arctic Planets.—The Moon and other Satellites.—Meteors and Comets: their Office in the Solar System.—Other Suns than Ours.—Of Minor Stars, and of the Distribution of Stars in Space.—The Nebulæ: are they External Galaxies?—Supervision and Control.

OUR PLACE AMONG INFINITIES. A Series of Essays contrasting our Little Abode in Space and Time with the Infinities around us. To which are added Essays on the Jewish Sabbath and Astrology. 12mo. Cloth, $1.75.

CONTENTS.—Past and Future of the Earth.—Seeming Wastes in Nature.—New Theory of Life in other Worlds.—A Missing Comet.—The Lost Comet and its Meteor Train.—Jupiter.—Saturn and its System.—A Giant Sun.—The Star Depths.—Star Gauging.—Saturn and the Sabbath of the Jews.—Thoughts on Astrology.

THE EXPANSE OF HEAVEN. A Series of Essays on the Wonders of the Firmament. 12mo. Cloth $2.00.

CONTENTS.—A Dream that was not all a Dream.—The Sun.—The Queen of Night.—The Evening Star.—The Ruddy Planet.—Life in the Ruddy Planet.—The Prince of Planets.—Jupiter's Family of Moons.—The Ring-Girdled Planet.—Newton and the Law of the Universe.—The Discovery of Two Giant Planets.—The Lost Comet.—Visitants from the Star Depths.—Whence come the Comets?—The Comet Families of the Giant Planets.—The Earth's Journey through Showers.—How the Planets Grew.—Our Daily Light.—The Flight of Light.—A Cluster of Suns.—Worlds ruled by Colored Suns.—The King of Suns.—Four Orders of Suns.—The Depths of Space.—Charting the Star Depths.—The Star Depths Astir with Life.—The Drifting Stars.—The Milky Way.

THE MOON: *Her Motions, Aspect, Scenery, and Physical Conditions.* With Three Lunar Photographs, Map, and many Plates, Charts, etc. 12mo. Cloth, $2.00.

CONTENTS.—The Moon's Distance, Size, and Mass.—The Moon's Motions.—The Moon's Changes of Aspect, Rotation, Libration, etc.—Study of the Moon's Surface.—Lunar Celestial Phenomena.—Condition of the Moon's Surface.—Index to the Map of the Moon.

LIGHT SCIENCE FOR LEISURE HOURS. A Series of Familiar Essays on Scientific Subjects, Natural Phenomena, etc. 12mo. Cloth, $1.75.

D. APPLETON & CO., 72 Fifth Avenue, New York.

D. APPLETON & CO.'S PUBLICATIONS.

CHAPTERS IN POLITICAL ECONOMY. By ALBERT S. BOLLES, Lecturer on Political Economy in the Boston University. Square 12mo. Cloth, $1.50.

CONTENTS.—The Field and Importance of Political Economy; The Payment of Labor; On the Increase of Wages; Effect of Machinery on Labor; On the Meaning and Causes of Value; A Measure of Value; Money and its Uses; Decline in the Value of Gold and Silver; The Money of the Future; The Good and Evil of Banking; The Financial Panic of 1873; Relation of Banks to Speculators; Influence of Credit on Prices; On Legal Interference with the Loan of Money, Payment of Labor, and Contracts of Corporations; Advantages of Exchange; Taxation.

PROTECTION VERSUS FREE TRADE. The Scientific Validity and Economic Operation of Defensive Duties in the United States. By HENRY M. HOYT. 12mo. Cloth, $2.00; paper, 50 cents.

The author of this work is well known as formerly Governor of Pennsylvania. He appears in this volume as a defender of protection, discussing the subject in a judicial spirit, with great fullness.

PROTECTION TO HOME INDUSTRY. Four Lectures delivered in Harvard University, January, 1885. By R. E. THOMPSON, A. M., Professor in the University of Pennsylvania. 8vo. Cloth, $1.00.

"In these lectures Professor Thompson has stated the essential arguments for protection so clearly and compactly that it is not strange that they have produced a deep impression. . . . The lectures as printed form a neat volume, which all fairly informed students may read with interest."—*Philadelphia Item.*

TALKS ABOUT LABOR, and concerning the Evolution of Justice between Laborers and Capitalists. By J. N. LARNED. 12mo. Cloth, $1.50.

The author's aim has been to find the direction in which one may hopefully look for some more harmonious and more satisfactory conjunction of capital with labor than prevails in our present social state, by finding in what direction the rules of ethics and the laws of political economy tend together.

HANDBOOK OF SOCIAL ECONOMY; or, The Worker's A B C. By EDMOND ABOUT. 12mo. Cloth, $2.00.

CONTENTS.—Man's Wants; Useful Things; Production; Parasites; Exchange; Liberty; Money; Wages; Savings and Capital; Strikes; Cooperation; Assurance, and some other Desirable Novelties.

New York: D. APPLETON & CO., 72 Fifth Avenue.

D. APPLETON & CO.'S PUBLICATIONS.

THE SUN. By C. A. YOUNG, Ph. D., LL. D., Professor of Astronomy in Princeton University. New and revised edition, with numerous Illustrations. 12mo. Cloth, $2.00.

"In this book we see a master's hand. Professor Young has no superiors, if he has rivals, among astronomers in this country. . . . 'The Sun' is a book of facts and achievements, and not a discussion of theories, and it will be read and appreciated by all scientific students, and not by them alone. Being written in untechnical language, it is equally adapted to a large class of educated readers not engaged in scientific pursuits."—*Journal of Education, Boston.*

"Professor Young's work is essentially a record of facts and achievements, rather than of theories and attempts at the interpretation of mysteries; yet the great questions still remaining to be answered are of course discussed, and in a masterly manner."—*Philadelphia Evening Bulletin.*

"It is one of the best books of popular science ever written, and deserves a host of readers."—*The Dial, Chicago.*

"You feel throughout that a master is leading you amid the intricacies and mazes of one of the most absorbing of studies. . . . Many a one whose views are hazy and dim will find here just that enlightenment, without an overburdened technicality, that will prove most useful."—*The Interior.*

THE STORY OF THE SUN. By Sir ROBERT S. BALL, F. R. S., author of "An Atlas of Astronomy," "The Cause of an Ice Age," etc. 8vo. Cloth, $5.00.

"Sir Robert Ball has the happy gift of making abstruse problems intelligible to the 'wayfaring man' by the aid of simple language and a few diagrams. Science moves so fast that there was room for a volume which should enlighten the general reader on the present state of knowledge about solar phenomena, and that place the present treatise admirably fills."—*London Chronicle.*

"As a specimen of the publisher's art it is superb. It is printed on paper which entices the reader to make marginal notes of reference to other books in his library, the type is large, the binding is excellent, and the volume is neither too large nor too small to handle without fatigue."—*New York Herald.*

AN ATLAS OF ASTRONOMY. By Sir ROBERT S. BALL, F. R. S., Professor of Astronomy and Geometry at the University of Cambridge; author of "Starland," "The Cause of an Ice Age," etc. With 72 Plates, Explanatory Text, and Complete Index. Small 4to. Cloth, $4.00.

"The high reputation of Sir Robert Ball as a writer on astronomy at once popular and scientific is in itself more than sufficient recommendation of his newly published 'Atlas of Astronomy.' The plates are clear and well arranged, and those of them which represent the more striking aspects of the more important heavenly bodies are very beautifully executed. The introduction is written with Sir Robert Ball's well-known lucidity and simplicity of exposition, and altogether the Atlas is admirably adapted to meet the needs and smooth the difficulties of young and inexperienced students of astronomy, as well as materially to assist the researches of those that are more advanced."—*London Times.*

New York: D. APPLETON & CO., 72 Fifth Avenue.

D. APPLETON & CO.'S PUBLICATIONS.

THE NATURAL HISTORY OF SELBORNE, AND OBSERVATIONS ON NATURE. By GILBERT WHITE. With an Introduction by John Burroughs, 80 Illustrations by Clifton Johnson, and the Text and New Letters of the Buckland edition. In two volumes. 12mo. Cloth, $4.00.

"White himself, were he alive to-day, would join all his loving readers in thanking the American publishers for a thoroughly excellent presentation of his famous book. . . . This latest edition of White's book must go into all of our libraries; our young people must have it at hand, and our trained lovers of select literature must take it into their homes. By such reading we keep knowledge in proper perspective and are able to grasp the proportions of discovery."—*Maurice Thompson, in the Independent.*

"White's 'Selborne' belongs in the same category as Walton's 'Complete Angler'; . . . here they are, the 'Complete Angler' well along in its third century, and the other just started in its second century, both of them as highly esteemed as they were when first published, both bound to live forever, if we may trust the predictions of their respective admirers. John Burroughs, in his charming introduction, tells us why White's book has lasted and why this new and beautiful edition has been printed. . . . This new edition of his work comes to us beautifully illustrated by Clifton Johnson."—*New York Times.*

"White's 'Selborne' has been reprinted many times, in many forms, but never before, so far as we can remember, in so creditable a form as it assumes in these two volumes, nor with drawings comparable to those which Mr. Clifton Johnson has made for them."—*New York Mail and Express.*

"We are loath to put down the two handsome volumes in which the source of such a gift as this has been republished. The type is so clear, the paper is so pleasant to the touch, the weight of each volume is so nicely adapted to the hand, and one turns page after page with exactly that quiet sense of ever new and ever old endeared delight which comes through a window looking on the English countryside—the rooks cawing in a neighboring copse, the little village nestling sleepily amid the trees, trees so green that sometimes they seem to hover on the edge of black, and then again so green that they seem vivid with the flaunting bravery of spring."—*New York Tribune.*

"Not only for the significance they lend to one of the masterpieces of English literature, but as a revelation of English rural life and scenes, are these pictures delightfully welcome. The edition is in every way creditable to the publishers."—*Boston Beacon.*

"Rural England has many attractions for the lover of Nature, and no work, perhaps, has done i's charms greater justice than Gilbert White's 'Natural History of Selborne.'"—*Boston Journal.*

"This charming edition leaves really nothing to be desired."—*Westminster Gazette.*

"This edition is beautifully illustrated and bound, and deserves to be welcomed by all naturalists and Nature lovers."—*London Daily Chronicle.*

"Handsome and desirable in every respect. . . . Welcome to old and young."—*New York Herald.*

"The charm of White's 'Selborne' is not definable But there is no other book of the past generations that will ever take the place with the field naturalists."—*Baltimore Sun.*

New York: D. APPLETON & CO., 72 Fifth Avenue.

LITERATURES OF THE WORLD.
Edited by EDMUND GOSSE,
Hon. M. A. of Trinity College, Cambridge.

ANCIENT GREEK LITERATURE. By Gilbert Murray, M. A., Professor of Greek in the University of Glasgow. 12mo. Cloth, $1.50.

"A sketch to which the much-abused word 'brilliant' may be justly applied. Mr. Murray has produced a book which fairly represents the best conclusions of modern scholarship with regard to the Greeks."—*London Times.*

"An illuminating history of Greek literature, in which learning is enlivened and supplemented by literary skill, by a true sense of the 'humanities.' The reader feels that this is no book of perfunctory erudition, but a labor of love, performed by a scholar, to whom ancient Greece and her literature are exceedingly real and vivid. His judgments and suggestions are full of a personal fresh sincerity; he can discern the living men beneath their works, and give us his genuine impression of them."—*London Daily Chronicle.*

"A fresh and stimulating and delightful book, and should be put into the hands of all young scholars. It will make them understand, or help to make them understand, to a degree they have never yet understood, that the Greek writers over whom they have toiled at school are living literature after all."—*Westminster Gazette.*

"Brilliant and stimulating."—*London Athenæum.*

"A powerful and original study."—*The Nation.*

"Mr. Murray's style is lucid and spirited, and, besides the fund of information, he imparts to his subject such fresh and vivid interest that students will find in these pages a new impulse for more profound and exhaustive study of this greatest and most immortal of all the world's literatures."—*Philadelphia Public Ledger.*

"The admirable perspective of the whole work is what one most admires. The reader unlearned in Greek history and literature sees at once the relation which a given author bore to his race and his age, and the current trend of thought, as well as what we value him for to-day.... As an introduction to the study of some considerable portion of Greek literature in English translations it will be found of the very highest usefulness."—*Boston Herald.*

"Professor Murray has written an admirable book, clear in its arrangement, compact in its statements, and is one, we think, its least scholarly reader must feel an instructive and thoroughly trustworthy piece of English criticism."—*New York Mail and Express.*

"At once scholarly and interesting.... Professor Murray makes the reader acquainted not merely with literary history and criticism, but with individual living, striving Greeks.... He has felt the power of the best there was in Greek life and literature, and he rouses the reader's enthusiasm by his own honest admiration."—*Boston Transcript.*

"Professor Murray has contributed a volume which shows profound scholarship, together with a keen literary appreciation. It is a book for scholars as well as for the general reader. The author is saturated with his subject, and has a rare imaginative sympathy with ancient Greece."—*The Interior, Chicago.*

"Written in a style that is sometimes spasmodic, often brilliant, and always fresh and suggestive."—*New York Sun.*

"Professor Murray's careful study will be appreciated as the work of a man of unusual special learning, combined with much delicacy of literary insight."—*New York Christian Advocate.*

D. APPLETON AND COMPANY, NEW YORK.

LITERATURES OF THE WORLD.

MODERN ENGLISH LITERATURE. By EDMUND GOSSE, Hon. M. A. of Trinity College, Cambridge. 12mo. Cloth, $1.50.

"Mr Gosse has been remarkably successful in bringing into focus and proportion the salient features of this vast and varied theme. We have read the book, not only with pleasure but with a singular emotion. . . . His criticism is generally sympathetic, but at the same time it is always sober."—*London Daily Chronicle.*

"Mr. Gosse's most ambitious book and probably his best. It bears on every page the traces of a genuine love for his subject and of a lively critical intelligence. Moreover, it is extremely readable—more readable, in fact, than any other single volume dealing with this same vast subject that we can call to mind. . . . Really a remarkable performance."—*London Times.*

"A really useful account of the whole process of evolution in English letters—an account based upon a keen sense at once of the unity of his subject and of the rhythm of its ebb and flow, and illumined by an unexampled felicity in hitting off the leading characteristics of individual writers."—*London Athenæum.*

"Probably no living man is more competent than Mr. Gosse to write a popular and yet scholarly history of English literature. . . . The greater part of his life has been given up to the study and criticism of English literature of the past, and he has a learned and balanced enthusiasm for every writer who has written excellently in English."—*London Saturday Review.*

"The bibliographical list is of extreme value, as is the bibliographical work generally. It is just one of these books which every reader will want to place among his working books."—*New York Times.*

"To have given in a moderate volume the main points in a literature almost continuous for five centuries is to have done a marvelous thing. But he might have done it dryly; he has made every sentence crisp and sparkling."—*Chicago Times-Herald.*

"A book which in soundness of learning, sanity of judgment, and attractiveness of manner has not been equaled by the work of any other author who has sought to analyze the elements of English literature in a concise and authoritative way."—*Boston Beacon.*

"Thoroughly enjoyable from first to last. It traces the growth of a literature so clearly and simply, that one is apt to underrate the magnitude of the undertaking. Mr. Gosse's charming personality pervades it all, and his happy manner illuminates matter that has been worked over and over until one might imagine all its freshness gone."—*Chicago Evening Post.*

"This is not a mere collection of brief essays on the merits of authors, but a continuous story of the growth of literature, of which the authors and their works are only incidents. The book is lucid, readable, and interesting, and a marvel of condensed information, without its seeming to be so. It can be read by nine out of ten intelligent people, not only without fatigue, but with pleasure; and when it is finished the reader will have a comprehensive and intelligent view of the subject which will not only enable him to talk with some ease and confidence upon the merits of the principal creators of English literature, but will also point the way to the right sources if he wishes to supplement the knowledge which he has derived from this book."—*Pittsburg Times.*

"That he has been a careful student, however, in many departments, the most unrelated and recondite, is evident on every page, in the orderly arrangement of his multitudinous materials, in the accuracy of his statements, in the acuteness of his critical observations, and in the large originality of most of his verdicts. He says things that many before him may have thought, though they failed to express them, capturing their fugitive expressions in his curt, inevitable phrases."—*N. Y. Mail and Express.*

D. APPLETON AND COMPANY, NEW YORK.

D. APPLETON AND COMPANY'S PUBLICATIONS.

LITERATURES OF THE WORLD.
Edited by EDMUND GOSSE,
Hon. M. A. of Trinity College, Cambridge.

FRENCH LITERATURE. By EDWARD DOWDEN, D. Litt., LL. D., D. C. L., Professor of English Literature in the University of Dublin. 12mo. Cloth, $1.50.

"Certainly the best history of French literature in the English language."—*London Athenæum.*

"This is a history of literature as histories of literature should be written. . . . A living voice, speaking to us with gravity and enthusiasm about the writers of many ages, and of being a human voice always. Hence this book can be read with pleasure even by those for whom a history has in itself little attraction."—*London Saturday Review.*

"The book is excellently well done; accurate in facts and dates, just in criticism, well arranged in method. . . . The excellent bibliography with which it concludes will be invaluable to those who wish to pursue the study further on their own lines."—*London Spectator.*

"Remarkable for its fullness of information and frequent brilliancy. . . . A book which both the student of French literature and the stranger to it will, in different ways, find eminently useful, and in many parts of it thoroughly enjoyable as well."—*London Literary World.*

"Professor Dowden is both trustworthy and brilliant; he writes from a full knowledge and a full sympathy. Master of a style rather correct than charming for its adornments, he can still enliven his pages with telling epigram and pretty phrase. Above all things, the book is not eccentric, not unmethodical, not of a wayward brilliance; and this is especially commendable and fortunate in the case of an English critic writing upon French literature."—*London Daily Chronicle.*

"A book readable, graphic, not overloaded with detail, not bristling with dates. . . . It is a book that can be held in the hand and read aloud with pleasure as a literary treat by an expert in style, master of charming words that come and go easily, and of other literatures that serve for illustrations."—*The Critic.*

"His methods afford an admirable example of compressing an immense amount of information and criticism in a sentence or paragraph, and his survey of a vast field is both comprehensive and interesting. As an introduction for the student of literature the work is most excellent, and for the casual reader it will serve as a compendium of one of the richest literatures of the world."—*Philadelphia Public Ledger.*

"Thorough without being diffuse. The author is in love with his subject, has made it a study for years, and therefore produced an entertaining volume. Of the scholarship shown it is needless to speak. . . . It is more than a cyclopædia. It is a brilliant talk by one who is loaded with the lively ammunition of French prose and verse. He talks of the pulpit, the stage, the Senate, and the *salon*, until the preachers, dramatists, orators, and philosophers seem to be speaking for themselves."—*Boston Globe.*

"Professor Dowden's book is more interesting than we ever supposed a brief history of a literature could be. His characterizations are most admirable in their conciseness and brilliancy. He has given in one volume a very thorough review of French literature."—*The Interior, Chicago.*

"The book will be especially valuable to the student as a safe and intelligible index to a course of reading."—*The Independent.*

D. APPLETON AND COMPANY, NEW YORK.

www.ingramcontent.com/pod-product-compliance
Lightning Source LLC
Chambersburg PA
CBHW030407230426
43664CB00007BB/785